TENTH ANNUAL SYMPOSIUM
ON
FRONTIERS OF ENGINEERING

NATIONAL ACADEMY OF ENGINEERING
OF THE NATIONAL ACADEMIES

THE NATIONAL ACADEMIES PRESS
Washington, D.C.
www.nap.edu

THE NATIONAL ACADEMIES PRESS   500 Fifth Street, N.W.   Washington, D.C. 20001

NOTICE: This publication has been reviewed according to procedures approved by a National Academy of Engineering report review process. Publication of signed work signifies that it is judged a competent and useful contribution worthy of public consideration, but it does not imply endorsement of conclusions or recommendations by the NAE. The interpretations and conclusions in such publications are those of the authors and do not purport to represent the views of the council, officers, or staff of the National Academy of Engineering.

Funding for the activity that led to this publication was provided by the Air Force Office of Scientific Research, Defense Advanced Research Projects Agency, Department of Defense–DDR&E-Research, National Aeronautics and Space Administration, Eastman Kodak Company, Microsoft Corporation, Cummins, Inc., ATOFINA Chemicals, Inc., Air Products and Chemicals, Inc., Dr. Ruth M. Davis, and other individual donors.

International Standard Book Number 0-309-09547-6 (Book)
International Standard Book Number 0-309-54784-9 (PDF)

Additional copies of this report are available from The National Academies Press, 500 Fifth Street, N.W., Lockbox 285, Washington, DC 20001; (800) 624-6242 or (202) 334-3313 (in the Washington metropolitan area); Internet, http://www.nap.edu.

Printed in the United States of America.

THE NATIONAL ACADEMIES
Advisers to the Nation on Science, Engineering, and Medicine

The **National Academy of Sciences** is a private, nonprofit, self-perpetuating society of distinguished scholars engaged in scientific and engineering research, dedicated to the furtherance of science and technology and to their use for the general welfare. Upon the authority of the charter granted to it by the Congress in 1863, the Academy has a mandate that requires it to advise the federal government on scientific and technical matters. Dr. Bruce M. Alberts is president of the National Academy of Sciences.

The **National Academy of Engineering** was established in 1964, under the charter of the National Academy of Sciences, as a parallel organization of outstanding engineers. It is autonomous in its administration and in the selection of its members, sharing with the National Academy of Sciences the responsibility for advising the federal government. The National Academy of Engineering also sponsors engineering programs aimed at meeting national needs, encourages education and research, and recognizes the superior achievements of engineers. Dr. Wm. A. Wulf is president of the National Academy of Engineering.

The **Institute of Medicine** was established in 1970 by the National Academy of Sciences to secure the services of eminent members of appropriate professions in the examination of policy matters pertaining to the health of the public. The Institute acts under the responsibility given to the National Academy of Sciences by its congressional charter to be an adviser to the federal government and, upon its own initiative, to identify issues of medical care, research, and education. Dr. Harvey V. Fineberg is president of the Institute of Medicine.

The **National Research Council** was organized by the National Academy of Sciences in 1916 to associate the broad community of science and technology with the Academy's purposes of furthering knowledge and advising the federal government. Functioning in accordance with general policies determined by the Academy, the Council has become the principal operating agency of both the National Academy of Sciences and the National Academy of Engineering in providing services to the government, the public, and the scientific and engineering communities. The Council is administered jointly by both Academies and the Institute of Medicine. Dr. Bruce M. Alberts and Dr. Wm. A. Wulf are chair and vice chair, respectively, of the National Research Council.

www.national-academies.org

ORGANIZING COMMITTEE

PABLO G. DEBENEDETTI (Chair), Class of 1950 Professor, Department of Chemical Engineering, Princeton University

KRISTI S. ANSETH, Tisone Professor of Chemical and Biological Engineering, Associate Professor of Surgery, and HHMI Assistant Investigator, Department of Chemical and Biological Engineering, University of Colorado, Boulder

DAVID BARAFF, Senior Animation Scientist, Pixar Animation Studios

DIANN E. BREI, Associate Professor, Department of Mechanical Engineering, University of Michigan

GRANT S. HEFFELFINGER, Deputy Director, Materials and Process Sciences Center, Sandia National Laboratories

CHRIS KYRIAKAKIS, Associate Professor, Department of Electrical Engineering, and Research Area Director, Sensory Interfaces, University of Southern California

MARY KAE LOCKWOOD, Aerospace Engineer, NASA Langley Research Center

DIMITRIOS MAROUDAS, Professor, Department of Chemical Engineering, University of Massachusetts

JOHN W. WEATHERLY, Ice Geophysicist, U.S. Army Cold Regions Research and Engineering Laboratory

Staff

JANET R. HUNZIKER, Program Officer
JENNIFER M. HARDESTY, Senior Project Assistant

Preface

In 1995, the National Academy of Engineering (NAE) initiated the Frontiers of Engineering Program, which brings together every year about 100 of the nation's future engineering leaders to learn about cutting-edge research and technical work in different fields of engineering. On September 9–11, 2004, NAE held its tenth U.S. Frontiers of Engineering Symposium at the Beckman Center of the National Academies in Irvine, California. Speakers were asked to prepare extended summaries of their presentations, which are reprinted in this volume. The intent of this book, and of the volumes that preceded it in the series, is to convey the revolutionary quality of this unique meeting and to highlight some exciting developments in engineering today.

GOALS OF THE FRONTIERS OF ENGINEERING PROGRAM

The practice of engineering is changing. Engineers must not only be able to thrive in an environment of rapid technological change and globalization, but they must also be able to work in interdisciplinary teams. The frontiers of engineering are at the intersections of engineering disciplines, and researchers and practitioners must be aware of developments and challenges in areas other than their own.

At the three-day Frontiers of Engineering Symposium, 100 of this country's best and brightest engineers, ages 30 to 45, have an opportunity to learn from their peers about work being done on the leading edges of engineering. It is hoped that the exchange of information on current developments in many fields of engineering will lead to insights that may be applicable in specific disciplines. In addition, the symposium gives engineers from a variety of institutions in academia, industry, and government, and from many different engineering disciplines, a chance to make contacts with and learn from individuals whom they would not meet in the usual round of professional meetings. This networking may lead to collaborative work and facilitate the transfer of new techniques and approaches across fields.

The number of participants at each meeting is limited to 100 to maximize opportunities for interactions and exchanges among the attendees, who are chosen through a competitive nomination and selection process. The choice of topics and speakers for each meeting is made by an organizing committee composed of engineers in the same 30- to 45-year-old cohort as the participants. Each year different topics are covered, and, with a few exceptions, different individuals participate.

The speakers at the Frontiers of Engineering Symposium must address a unique challenge—to communicate the excitement of their work to a technically sophisticated, but nonspecialist audience. To achieve the objectives of the meeting, speakers are asked to provide a brief overview of their fields and to address several aspects of their topics: a description of the frontiers in the field, the experiments, prototypes, and design studies that have been completed or are in progress, new tools and methodologies, limitations on advances and controversies, and the theoretical, commercial, societal, and long-term significance of the work.

THE 2004 SYMPOSIUM

The four broad topics for the 2004 meeting were engineering for extreme environments, designer materials, multiscale modeling, and engineering and entertainment. In the Engineering for Extreme Environments session, speakers addressed the challenges of designing devices and developing technologies for polar climates, deep oceans, high-current rivers, nuclear power plants, and the lunar and Martian surfaces. Talks covered the deployment of robots on the Antarctic plateau for long- or short-term observation; state-of-the-art modeling and simulation techniques used in constructing the Tacoma Narrows Bridge, a nuclear-waste processing plant, and the Chernobyl New Safe Confinement Structure; the landing of robotic systems on the surface of Mars; and the challenges of accessing the lunar poles for human exploration missions. In the Designer Materials session, speakers described the elegant and rationale design of highly functional materials with particular qualities. Three specific topics were addressed: thin-film active materials, which are defined by dimensions on the order of microns and present unique physical coupling phenomena; hybridized materials with performance-tailored functions; and small-diameter tissue-engineered vascular grafts, which could improve the quality of life for people with vascular disease by preventing thrombosis and improving graft mechanical properties. The Multiscale Modeling session was a departure from past Frontiers sessions in that it focused on an engineering tool rather than a particular engineering field. However, because multiscale modeling and simulation are used in many areas of science and engineering research, the presentations were of interest to everyone. The four speakers focused on different aspects of multiscale modeling, including the coupling of modeling and simulation methods across time and length scales for specific applications, such as the processing of engineering materials (e.g., semiconductors, metals, and polymers); biological applications; the health sciences; and climatology. The final session, Engineering and Entertainment, focused on three areas—picture, sound, and actors. The presentations covered new computer-graphics techniques that make the visual elements of film practically indistinguishable from reality, technologies for spatial sound reproduction and the prospects for individualized binaural sound, and socially intelligent personal-service robots that can not only entertain but can also participate in people's daily lives.

NAE is deeply grateful to the following organizations for their support of the Tenth Annual Symposium on Frontiers of Engineering: Air Force Office of Scientific Research, Defense Advanced Research Projects Agency, U.S. Department of Defense-DDR&E Research, National Aeronautics and Space Administration, Eastman Kodak, Microsoft Corporation, ATOFINA Chemicals, Inc., Cummins, Inc., Air Products and Chemicals, Inc., and Dr. Ruth M. Davis and other individual donors. NAE would also like to thank the members of the Symposium Organizing Committee chaired by Professor Pablo Debenedetti, for planning and organizing the event (see p. iv).

Contents

MULTISCALE MODELING

ENGINEERING AND ENTERTAINMENT

APPENDIXES

ENGINEERING FOR EXTREME ENVIRONMENTS

Introduction

MARY KAE LOCKWOOD
NASA Langley Research Center
Hampton, Virginia

JOHN W. WEATHERLY
U.S. Army Cold Regions Research
and Engineering Laboratory
Hanover, New Hampshire

Designing for extreme environments presents unique challenges for engineers. Materials and devices must be able to function in remote locations and under very harsh conditions, such as extreme natural environments (e.g., polar regions, deep oceans, and high-current rivers), man-made environments (e.g., the inside of a nuclear facility or a space vehicle), and in space (e.g., on the surface of the Moon or Mars). In addition, engineering for extreme environments involves many uncertainties—about specific environmental conditions, about the physics under extreme environmental conditions, about the accuracy of laboratory simulations, and about meeting difficult schedule and cost requirements.

The Antarctic plateau is a unique environment for studying the upper atmosphere at very high magnetic latitudes and a range of longitudes. One can envision a "network" of robots, with instruments secured, being deployed from the South Pole station to locations on a geomagnetic grid on the surface of the plateau for long- or short-term observation. The robots might be retrieved or repositioned through iridium-based communication. Very large-scale arrays of robots can provide ground-based, distributed, mobile "antennas" as an alternative, real-time communication and high-bandwidth data-transmission link to the iridium system. With a sufficiently stable robotic platform, a wide range of instruments, from seismometers to radio receivers, could be deployed. Robot networks would enable field workers to take advantage of the value of large sensor arrays for geophysical sciences and ecology. Robots would also be much less expensive than stationary observatories (which are ideal for long-term observation but prohibitively expensive for the deployment of very large-scale instrument networks).

The construction of the Tacoma Narrows Bridge is an example of engineering in an extreme environment that presented significant technical challenges and strict schedule constraints. A complex, cable-supported anchoring system was necessary to stabilize massive floating concrete caissons 80 × 130 × 150 feet deep. The project required accurate predictions of large current-driven, time-dependent loads on the floating, but stabilized caissons. However, no relevant field measurements for similar structures or experimental scale models were available. Researchers, therefore, used state-of-the-art computational fluid-dynamics methods combined with an aggressive team approach to meet the extreme technical challenges and the tight schedule constraints. A similar approach, with specifically tailored computational fluid-dynamics analyses, was used for the design of a pulsed-jet mixer system for the construction of the largest nuclear-waste processing plant in the United States.

In January 2004, two Mars explorer rover (MER) spacecraft—Spirit and Opportunity—were successfully landed on Mars. The entry, descent, and landing systems were designed to slow the spacecraft from a velocity of almost 6 km/sec at the edge of the Martian atmosphere to nearly zero velocity approximately 12 m above the Martian surface to enable the safe release of the airbag/landing system. The MER landings followed earlier successes—the landings of two Viking missions and the Mars Pathfinder. Future Mars landings, such as the Mars Science Laboratory mission slated for launch in 2009, will include a sky-crane landing system that can set down larger mass robotic systems on the Martian surface.

The recent announcement by President Bush of a new vision for space exploration, including human travel to the Moon, Mars, and beyond, will present many additional challenges in the design, development, and testing of systems for human missions in extreme environments. Researchers are already assessing systems for supplying astronauts with artificial gravity to minimize bone decalcification and provide an accommodating environment for the long transit from Earth to Mars. Investigations of previously unexplored regions of the Moon, such as the polar caps, are also being considered. Because temperatures in those areas have remained below 40K for millions of years, the environment has remained unchanged for millions of years.

These are only a few examples of the risks and rewards of engineering for extreme environments. Innovative solutions have been, and will continue to be, applied to some extremely challenging problems.

Cool Robots:
Scalable Mobile Robots for Instrument Network Deployment in Polar Climates

Laura R. Ray, Alexander D. Price,
Alexander Streeter, and Daniel Denton
Thayer School of Engineering
Dartmouth College
Hanover, New Hampshire

James H. Lever
U.S. Army Cold Regions Research and Engineering Laboratory
Hanover, New Hampshire

The Antarctic plateau is a unique location to study the upper atmosphere at high magnetic latitudes because it provides a stable environment for sensitive instruments that measure interactions between the solar wind and the earth's magnetosphere, ionosphere, and thermosphere. Existing stations on the edge of the continent and at the South Pole, and six low-power (50 W) automatic geophysical observatories, have demonstrated the value of distributed, ground-based observations of solar-terrestrial physics. Increasing the spatial density of these observations offers great scientific opportunities. A National Research Council report, *The Sun to the Earth and Beyond*, emphasizes the need for mobile instrument networks. The report recommends "comprehensive new approaches to the design and maintenance of ground-based, distributed instrument networks, with proper regard for the severe environments in which they must operate" (NRC, 2002).

This paper describes scalable mobile robots that can enable the deployment of instrument networks in Antarctica. The drivetrain, power system, chassis, and navigation algorithms can be scaled for payloads of roughly 5 to 25 kilograms. One can envision deploying robots from the South Pole to locations on the plateau for long-term or short-term observation and retrieving or repositioning the network through iridium-based communication. Potential missions include: deploying arrays of magnetometers, seismometers, radio receivers, and meteorological instruments; measuring disturbances in the ionosphere through synchronization of Global Positioning System (GPS) signals; using ground-penetrating

5

radar (GPR) to survey crevasse-free routes for field parties; and conducting gla-
ciological surveys with GPR. Robot arrays could also provide high-bandwidth
communications links and power systems for field scientists.

Based on this concept, a single robot is under construction that would carry
a triaxial fluxgate magnetometer, an iridium modem, and a modest set of weather
instruments. The magnetometer will be a payload test case. Magnetometer arrays
are already used in low latitudes and midlatitudes, but polar regions provide
unique windows for observing the effects of solar wind on the Earth's magnetosphere. With mobile networks, sensor locations could potentially be tuned to
events in the magnetosphere. Also, synchronized data from polar networks could
potentially discover spatial characteristics of narrow-band spectral features in
geomagnetic field data, identify magnetospheric boundaries, and refine models
accordingly (Lanzerotti et al., 1999).

The deployment of a remote observatory on the Antarctic plateau via transport and small aircraft is expensive and entails dangerous takeoffs and landings
at remote sites. For large-scale, widely distributed (>500 km radius) networks,
relatively low-cost mobile robots could substantially reduce per-instrument deployment costs. Semiautonomous network deployment would also free limited
aircraft and human resources for other missions.

The harsh weather of polar environments, long-range requirements, navigation issues, and variable terrain pose significant design challenges for inexpensive unmanned vehicles. Instruments would be deployed for long periods of time
in drifting snow and will require a stable environment with low vibration and
electromagnetic noise. Robots and deployed sensors should be retrievable with
high reliability to minimize environmental impact and cost. In this paper, related
robotics research is summarized, including mobile robot design concepts for
polar environments, technical challenges associated with their development, and
enabling technologies for cost-effective mobile robots.

STATE-OF-THE-ART ROBOTS FOR EXTREME ENVIRONMENTS

NOMAD, a gasoline-powered robot for polar and desert environments, was
developed at Carnegie Mellon University (Apostolopoulos et al., 2000; Carnegie
Mellon University, 2004b). NOMAD, which is 2.4 × 2.4 × 2.4 m in size and
weighs 725 kg (Figure 1), can travel up to 50 centimeters per second (cm/s) and
deploy instruments, such as a magnetometer. In 1997, NOMAD executed its first
mission in the Atacama Desert of southern Chile, traversing 223 km through
teleoperation. Subsequently, NOMAD successfully found and classified five in-
digenous meteorites on an Antarctic mission. For our purposes, however,
NOMAD's large size, high cost, and nonrenewable fuel are disadvantages. In
addition, its deployment experience suggests that navigation cameras may work
poorly in polar climates due to reflection of sunlight off of the snowfield.

FIGURE 1 NOMAD gasoline-powered robot. Source: Carnegie Mellon University, 2004b. Reprinted with permission.

Spirit and Opportunity are Mars exploration rovers developed by the National Aeronautics and Space Administration Jet Propulsion Laboratory. Each 2.3 × 1.6 × 1.5-m rover weighs 174 kg and has a top speed of 5 cm/s (NASA/ JPL, 2004). The power source for Mars rovers, a multipanel solar array and two rechargeable lithium-ion batteries, enables the rover to generate 140 W of power for four hours per sol, when the panels are fully illuminated. The warm electronics box, which contains the batteries, electronics, and computer, can only operate in the range of –40°C to +40°C. Gold-painted, insulated walls, solid silica aerogel, thermostat and heaters, and a heat rejection system protect the body from the 113°C temperature swing during the Martian day. The payload includes a panoramic camera, a miniature thermal emission spectrometer, a Mössbauer spectrometer, an alpha-particle X-ray spectrometer, and a rock abrasion tool. Each of the six wheels is driven by its own in-wheel motor, and the two front and two rear wheels have steering motors for point turns. The rovers are well suited for their mission on Mars but are too expensive for the deployment of instrument networks.

Hyperion (Figure 2), designed for sun-synchronous exploration, is a 157-kg, 2 × 2.4 × 3-m vehicle that includes a 3.45-m^2 nearly vertical solar panel; its maximum speed is 30 cm/s (Carnegie Mellon University, 2004a; Wettergreen et al., 2001). The chassis is intentionally simple—a 1.5 N-m, 150 W brushless DC motor combined with a harmonic drive for an 80:1 reduction ratio drives a wheel through a bicycle chain. A passively articulated steering joint provides two free rotations, enabling moderate maneuverability and mechanical simplicity. This

FIGURE 2 Hyperion rover. Source: Carnegie Mellon University, 2004a. Reprinted with permission.

design has many appealing features for instrument-network deployment, including renewable fuel, simplicity, and potentially low cost.

Commercial all-terrain robots made by iRobot and ActivMedia weigh 39 to 100 kg and carry payloads of 7 to 100 kg. Powered by two DC servomotors and a 4-wheel differential-drive system, these battery-operated robots run for 2 to 6 hours at speeds between 1 and 2 m/s. With no navigation instruments or software, these robots cost from $7,000 to $22,000. They could potentially be retrofitted for solar-power operation, but they are not designed for low-temperature operation, and the solar panels alone could comprise the entire payload budget.

COOL ROBOT CONCEPT

Our robot is designed to operate in interior Antarctica, which is characterized by low snowfall, moderate winds, and extreme cold. We envision networks of robots, guided by GPS and onboard sensors, that are launched and retrieved from the South Pole Station during the austral summer. Key design issues are outlined below.

Figure 3 shows a satellite photo of Antarctica. The vast central plateau covers more than five million square kilometers of relatively flat, crevasse-free terrain. A second large area of operation is the Ross Ice Shelf. Generally, Antarctic snowfields consist of dense, windblown snow. Aside from wind-sculpted sastrugi, dune-like features that are identifiable on satellite imagery, there are few obstacles. The central plateau receives less than 50 mm precipitation (<500 mm snowfall) in an average year. During summer months at the South Pole,

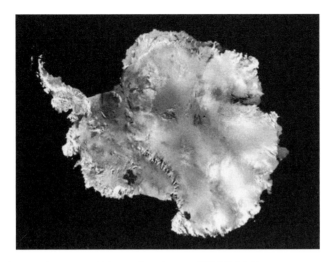

FIGURE 3 Satellite photo of Antarctica. Source: USGS, 2004.

wind speed averages 2 m/s (Valenziano and Dall'Oglio, 1999), and the five-year maximum speed is 20.5 m/s (CMDL, 2004). The average daily temperature is –20°C to –40°C.

An Antarctic robot must traverse firm snow and occasional softer drifts, sustain mobility in windy conditions, have a minimal environmental impact, and operate in temperatures down to –40°C. We envision a lightweight, solar-powered, wheeled robot that can be transported in a Twin Otter aircraft and is capable of traversing 500 km within two weeks. After reaching a target location, the robot could collect data over a period of two to three months before returning to the South Pole for the winter. The design includes a low center-of-gravity vehicle with four direct-drive brushless electric motors, an enclosed thermally controlled volume for instrumentation and batteries, and a solar panel "box" for renewable energy. Table 1 provides design specifications for a wheeled robot and a price point for economic viability for the deployment of networks of such robots.

Motion resistance in snow is attributable to sinkage and varies with the firmness of the snow pack immediately in front of the wheel, the length of the tire in contact with the snow, and the width of the tire (Richmond et al., 1995). Given the target ground pressure (< 20 kPa) in the dense snow of the Antarctic plateau, sinkage should be small. The estimated total resistance of 0.25 for a 90-kg vehicle will require a net traction force of 221 N. Travel of 500 km in two weeks will require an average speed of 0.41 m/s, with an average power requirement of 90 W and maximum power of 180 W for the top speed of 0.8 m/s. Allowing up to 40 W for housekeeping power and power-system efficiencies, the target power budget is approximately 220 W.

Table 1 Robot Specifications

Maximum speed	≥ 0.80 m/s
Mass (excluding payload)	≤ 75 kg
Payload mass	≥ 15 kg
Ground pressure	≤ 20 kPa
Operating temperature range	0 C to –40 C
Dimensions	≤ 1.4 × 1.15 × 1-m
Cost	≤ $20000

Despite a harsh climate and low sun angles, Antarctica is an ideal place for a solar-powered robot. The summer sun provides a 24-hr energy source, and the central plateau receives scant precipitation and infrequent fog. The Antarctic plateau is nearly completely covered in snow, with albedo averaging 95 percent across visible and ultraviolet wavelengths (Grenfell et al., 1994) and fairly uniform scattering in all directions (Warren et al., 1998). The high altitude and dry air block less incoming radiation, and there is a small benefit due to the proximity of the Earth to the sun during the summer. The sun remains at approximately the same elevation throughout the day (especially near the South Pole), resulting in relatively constant energy input. With low elevation angles and significant reflected solar energy, nearly vertical solar panels will be optimal. Also, the efficiency of solar cells increases as temperature decreases.

Average horizontal insolation data for 2002 (Figure 4) show a range of horizontal irradiance of 300 to 500 W/m^2 at the South Pole (CMDL, 2004). Adjusting for sun elevation angle gives a net insolation between 800 W/m^2 and 1,200 W/m^2, which is consistent with earlier studies (Hanson, 1960). At other locations in Antarctica where cloud cover and fog are more frequent, the average insolation is about half this, but the sunny days are almost as bright. For comparison, consider a clear winter day in New England. At a sun elevation of 20 degrees, the total insolation is between 600 and 800 W/m^2. Table 2 summarizes insolation data at various locations on the Antarctic continent and elsewhere. The average solar energy input during the November to February operating window is approximately 1,000 W/m^2, with an average sun elevation of about 20 degrees.

To determine the optimum size of the solar panel, we developed a model to predict power as a function of solar insolation, sun elevation, and azimuth for solar panels in a snowfield. The model assumes diffuse reflection from the snow at a specified albedo. We validated the model using data collected with a commercial 20-W panel during January-February 2004 in Hanover, New Hampshire. Figure 5 shows the resulting robot design concept—a wheeled chassis enclosed by a five-panel box—along with predicted panel capacities extrapolated from the model for nominal Antarctic solar radiation, 20 degree sun elevation, and 90 percent albedo. The panel outputs are reported as a percentage of their standard

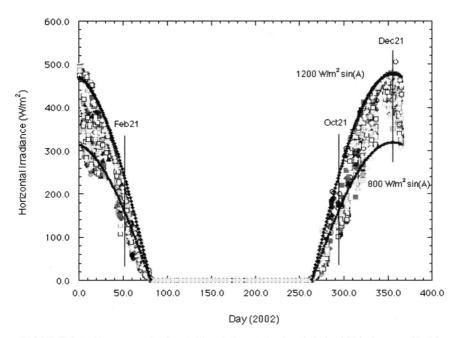

FIGURE 4 Daily average horizontal insolation at the South Pole, 2002. Source: CMDL, 2004.

Table 2 Insolation for Various Sun Conditions

Condition/Location	Nominal Insolation
Max. for Antarctica, continent	1200 W/m^2
Avg. for South Pole, Nov-Feb	1000 W/m^2
Avg. for South Pole, at solstice	1100 W/m^2
Avg. for south polar plateau	> 800 W/m^2
Avg. for Ross Ice Shelf	400 W/m^2
Jan 2004 measurements, Hanover, NH	660 W/m^2

capacities (rated at 1,000 W/m^2 insolation). The front panel (directly facing the sun) has a capacity of 128 percent (more than 100 percent due to reflected energy). Significantly, the top and side panels contribute nearly as much power as the panel facing the sun. Even the back panel receives substantial radiation because the robot's shadow is not as large as the area of snow that reflects light to the panel.

Enabling technology for the robot is the affordable, 20 percent efficient, A-300 solar cell by Sunpower, Inc., which became available in 2003. Figures 6 and

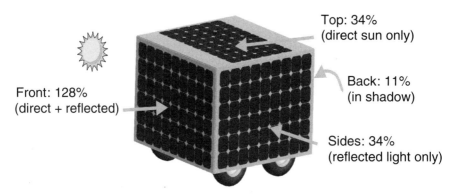

FIGURE 5 Panel power capacities in nominal Antarctic sun.

7 show predicted power available to the motors for a robot using 54 of these cells per panel (each cell is 12.5 × 12.5 cm) at 1,000 W/m² insolation and 90 percent albedo. The resulting robot will fit in the Twin Otter cargo bay. These results include efficient maximum-power-point-tracking (MPPT) circuits for each panel and subtract housekeeping power. The robot can drive at full speed even in below-average insolation. Under minimal insolation, the robot still has enough power to drive slowly or charge the batteries and drive in short bursts on battery power. Diffuse incoming radiation—light scattered by the atmosphere—is an unmodeled benefit because diffuse light is expected to be received perpendicular

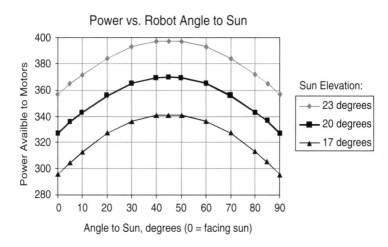

FIGURE 6 Power availability as a function of sun elevation angle and the robot azimuth angle.

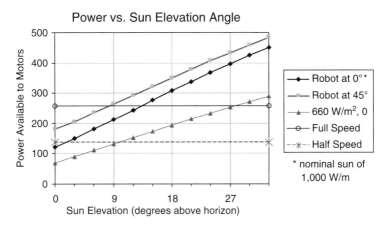

FIGURE 7 Power available and required at various sun angles.

to the panels from all directions. The total diffuse radiation is expected to be 50 to 100 W/m^2 providing 40 to 80 W, enough to drive the robot in bursts and maintain instrument operation (Hanson, 1960).

A positive feature of the deployment of instruments on the Antarctic plateau is relatively flat, straight paths on dense snow, which minimizes planning the complexity of the path. One of the most promising navigation architectures for such paths is "mixed-mode" operation, which mimics human behavior (e.g., in hiking a known path over a long distance) (Simmons et al., 1995). The global objective is to stay on the path, but a local mode, in which the hiker goes around unanticipated objects (e.g., downed trees), is in force for short periods, after which the hiker returns to the path. In the initial stages of research, global navigation is being used, primarily through GPS and speed control, with sensors to detect unbalanced wheel speeds and hence potential problems with traction, and low-bandwidth path-correction algorithms to reduce "dither" around the path. In the traversal of long distances, GPS-induced path deviations are tolerable. Traction control can be layered onto the basic global path-following algorithm, along with sensors for tilt and wheel slip. Local-mode navigation would be invoked if sensors detect extreme tilting or slippage.

Navigation and motion control also depend on the power system. The robot will move along the path only when adequate solar power is available. In cloudy conditions, the robot will move under battery power if necessary to prevent drifting in, and in windy conditions, the robot may face a direction that minimizes drag and the potential for tipping. We do not anticipate the robot having a vision system because sastrugi are visible on satellite imagery and can be avoided through route selection. The lack of contrast in snow-covered terrain would make onboard navigation challenging and potentially expensive.

DESIGN EMBODIMENT AND ENABLING TECHNOLOGIES

We have attempted to minimize vehicle mass by using stiff honeycomb composites for the solar panels and chassis and custom-designed wheel rims and hubs. The enabling technologies and cost-design trade-offs and an estimate of the costs of parts and the mass for the prototype robot are highlighted below.

Because of their newness, complete panels for A-300 solar cells are not yet available. Moreover, traditional panel construction with its steel backing is not viable for the robot. Thus, we will construct the solar panels in house, using quarter-inch honeycomb sandwich panels (Nomex core with fiberglass facing). The cell will be encapsulated in silicone. Similar honeycomb composite will be used to make the chassis box (Figure 8). The construction and joinery of honeycomb panels are mature technologies in the aerospace field, and these materials supply area densities of 1.4 kg/m² for solar panels and 2.5 kg/m² for chassis walls.

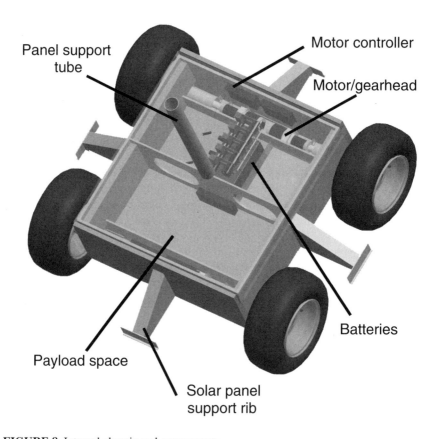

Panel support tube

Motor controller

Motor/gearhead

Payload space

Solar panel support rib

Batteries

FIGURE 8 Internal chassis and components.

The wheels are sized to provide adequate ground clearance and low rolling resistance. We considered many different tires and trade-offs between traction, weight, pressure, and rolling resistance. Taking into account availability and cost, a 16 × 6–8 ATV tire, which has low mass and excellent traction, was selected. Low-mass rims and hubs are not available for ATV tires, however, so these were custom designed and machined in house to meet the strength and deflection requirements for a 90-kg robot. For a mass-produced robot, rims and hubs could be cast or stamped at low cost.

High-efficiency, brushless motors with 90 percent efficient gear trains and lubricant for operation at –50°C drive the wheels directly. Each motor has a controller that can be configured for speed or torque control. We tested a single motor-gearbox combination in a cold room in a box insulated as configured on the robot. The results showed that both long-term operation and start-stop operation would be efficient, and controllability would be maintained at cold temperatures.

The power-system architecture includes three lithium-ion batteries in series and five solar panels, each of which can operate under varying insolation and temperature conditions. To deliver power to a common power bus, each panel requires an MPPT, a device that allows each panel to operate at the bus voltage established by the batteries while meeting power demands for the motors. Custom-designed MPPTs that weigh less than 250 grams and have better than 97 percent efficiency have been designed and constructed.

We estimate that the five-panel robot, without payload, will weigh ~73 kg, with a total material cost of less than $15,000. The design is relatively insensitive to payload up to about 20 kg. The robot will be equipped with instruments to assess power-input and mobility models during field trials in Antarctica, anticipated for the 2005–2006 austral summer.

CONCLUSION

Solar-powered mobile robots for operation on the Antarctic plateau are feasible in terms of power availability, mechanical design, and power-system design. Waypoint navigation on the relatively obstacle-free plateau through GPS can provide long-distance travel appropriate to the scientific missions envisioned. Mobile robots capable of reliable, long-term operation on the Antarctic plateau can potentially advance scientific research through instrument deployment, mapping, and the provision of portable mobile power.

REFERENCES

Apostolopoulos, D., M.D. Wagner, B. Shamah, L. Pedersen, K. Shillcutt, and W.L. Whittaker. 2000. Technology and field demonstration of robotic search for antarctic meteorites. International Journal of Robotics Research 19(11): 1015–1032.

Carnegie Mellon University. 2004a. Hyperion: Sun Synchronous Navigation. Available online at: <*http://www.ri.cmu.edu/projects/project_383.html*> (June, 2004).

Carnegie Mellon University. 2004b. Robotic Antarctic Meteorite Search: The NOMAD Robot. Available online at: <*http://www.frc.ri.cmu.edu/projects/meteorobot/Nomad/Nomad.html# Mechanical*> (June, 2004).

CMDL (Climate Monitoring and Diagnostic Laboratory). 2004. Daily Average Horizontal Insolation at the South Pole. Available online at: <*http://www.cmdl.noaa.gov/ info/ftpdata.html*> (March, 2004).

Grenfell, T.C., S.G. Warren, and P.C. Mullen. 1994. Reflection of solar radiation by the Antarctic snow surface at ultraviolet, visible, and near-infrared wavelengths. Journal of Geophysical Research 99(D9): 18-669–18-684.

Hanson, K.J. 1960. Radiation studies on the south polar snowfield. IGY Bulletin 31: 1–7. Washington, D.C.: National Academy of Sciences.

Lanzerotti, L., A. Shona, H. Fukunishi, and C.G. Maclennan. 1999. Long-period hydromagnetic waves at very high geomagnetic latitudes. Journal of Geophysical Research 104(A12): 8, 423.

NASA/JPL (National Aeronautics and Space Administration Jet Propulsion Laboratory). 2004. Spacecraft: Surface Operations: Rover. Available online at: <*http://www.marsrovers.jpl.nasa.gov/ mission/spacecraft_rover_energy.html*> (June, 2004).

NRC (National Research Council). 2002. The Sun to the Earth and Beyond: A Decadal Research Strategy in Solar and Space Physics. Washington, D.C.: National Academy Press.

Richmond, P.W., S.A. Shoop, and G.L. Blaisdell. 1995. Cold Regions Mobility Models. CRREL Report 95-1. Hanover, N.H.: Cold Regions Research and Engineering Laboratory. Available online at: <*http://www.stormingmedia.us/82/ 8273/ A827392.html*>.

Simmons, R., E. Krotkov, L. Chrisman, F. Cozman, R. Goodwin, M. Hebert, L. Katraqadda, S. Koenig, G. Krishnaswamy, Y. Shinoda, W. Whittaker, and P. Klarer. 1995. Experience with Rover navigation for lunar-like terrains. Journal of Engineering and Applied Science 1: 441–446.

USGS (U.S. Geological Survey). 2004. Satellite Image Map of Antarctica. Available online at: <*http://terraweb.wr.usgs.gov/TRS/projects/Antarctica/AVHRR.html*> (June, 2004).

Valenziano, L., and G. Dall'Oglio. 1999. Millimetre astronomy from the high Antarctic plateau: site testing at Dome C. Publications of the Astronomical Society of Australia 16: 167–174.

Warren, S.G., R.E. Brandt, and P. O'Rawe Hinton. 1998. Effect of surface roughness on bidirectional reflectance of Antarctic snow. Journal of Geophysical Research 103(E11): 25-789–25-807.

Wettergreen, D., B. Shamah, P. Tompkins, and W.L. Whittaker. 2001. Robotic Planetary Exploration by Sun-Synchronous Navigation. In Proceedings of the 6th International Symposium on Artificial Intelligence, Robotics and Automation in Space (i-SAIRAS 2001), Montreal, Canada, June 2001.

The Role of Modeling and Simulation in Extreme Engineering Projects

JON BERKOE
Bechtel National, Inc.
San Francisco, California

Bechtel has constructed a vast array of major plants and infrastructure projects around the world. In the past few years, advanced-technology tools, such as simulations, have increasingly been used to address logistical barriers, schedule constraints, environmental factors, and risk in support of the design and engineering in many projects.

Tools that specialize in modeling physical environments and conditions are particularly useful for many complex projects (e.g., plants and infrastructure). A multidisciplinary approach to analysis and visualization can reduce risk and save money by developing models that simulate the physical environment a component or system may encounter before substantial time and/or money has been invested in the project. In particular, models can be used to investigate safety implications of complex, off-normal conditions that cannot be easily evaluated by the project engineering team.

TACOMA NARROWS BRIDGE MOORING SYSTEM

The towing last July of the first 14,000-ton (12,700-tonne) Tacoma Narrows Bridge caisson to its moored position at the east end of Puget Sound was a historic sight witnessed by hundreds of people. What they didn't see, however, was the mooring system that restrained the caisson and its west-end counterpart until they took their underwater positions 60 to 70 feet (18 to 21 meters) below the Narrows mud line.

The caisson-mooring system was the result of nine months of design work and close coordination among several engineering teams and individual experts. The system was designed to handle the tidal conditions—8-knot currents and 17-foot (5-meter) tidal fluctuations—while responding to the volatile hydrodynamic zone in the Narrows created by vortex shedding from the existing bridge foundations.

The uniqueness of the project presented the mooring design team with several challenges. Here are some highlights:

1. The new structure will be the longest suspension bridge built in the United States in 40 years. Physical tests of the mooring system were not feasible at a scale that would effectively model the forces on the caissons in the Tacoma Narrows environment.

2. The new bridge foundations will be constructed approximately 80 feet (24 meters) from the existing bridge foundations. (The existing bridge will continue to operate after the new bridge is constructed.) This created extremely tight constraints on acceptable movement of the caissons.

3. The caisson and mooring system will be subject to extreme environmental conditions, particularly tidal fluctuations, in addition to storm surges, wind, and waves. Designing a mooring system with allowances for wide variations in environmental conditions and differing effects on each caisson was perhaps the biggest challenge.

Computational fluid dynamics (CFD)—a computer-based tool for simulating the behavior of systems involving fluid flow, heat transfer, and related physical processes—was used to predict the time-varying loads and moments on the new bridge caissons caused by current flows in the Narrows. To accomplish this, the bathymetry of the riverbed and the designs for the existing and new bridge piers had to be combined into one model; software that used an advanced finite-element-based solver and large eddy simulation proved to be fast and stable. The model showed that CFD results for the loads on the caissons agreed very well with experiment-scale model data from tests carried out at HR Wallingford in the United Kingdom. Using CFD modeling, the designers of the bridge caissons were able to assess risk and gain confidence in design margin, particularly for the "untested" west pier. Based on results of the analytical and physical modeling, Tacoma Narrows constructors selected a two-tiered anchor system with 16 anchors on each level (Figure 1).

CONSTRUCTION OF THE LARGEST NUCLEAR-WASTE PROCESSING PLANT IN THE UNITED STATES

Beside the Columbia River in Washington, 53 million gallons of radioactive waste (60 percent of the nation's radioactive waste) is stored in 177 underground

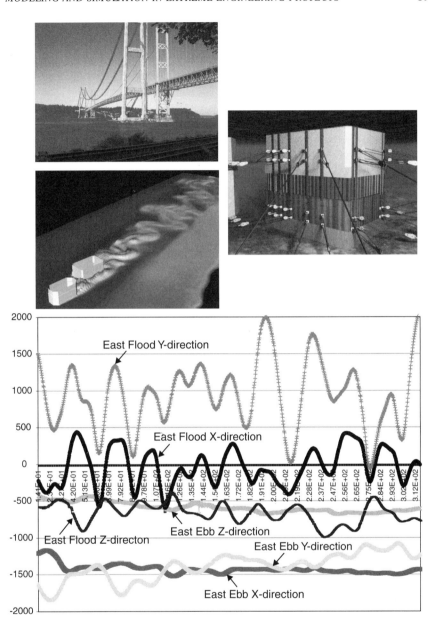

FIGURE 1 (Top to bottom) Rendering of the new Tacoma Narrows Bridge positioned next to the existing bridge. Caisson anchor system during installation. CFD analysis plot showing turbulence around two bridge piers subjected to current flow. Plot of the time-varying load components subjected to the caisson during ebb and flood flows. Source: Bechtel National, Inc.

tanks. This waste is a product of 50 years of plutonium production for national defense. The U.S. Department of Energy has commissioned the construction of a vast waste-treatment plant (WTP) to convert this waste into stable glass. The waste in these underground storage tanks is a combination of sludge, slurry, and liquid, which will be transported to a pretreatment facility to be processed in various vessels in preparation for vitrification.

Part of the engineering challenge is designing a system that can keep solids in continuous suspension during processing while minimizing the risk of mixing-system failure and eliminating the need for maintenance. Pulse-jet mixers (PJMs) are an integral part of the black-cell conceptual design. Black cells—sealed areas of the plant that no human will ever enter—are designed to require no maintenance, no equipment replacement, and no repairs. PJMs are air-driven pumps installed inside stainless-steel tanks that process radioactive waste. Because they have no moving parts, they never require maintenance or replacement. The mixers agitate the radioactive waste and keep it homogenous, which is necessary to achieve the correct blend of waste fed to the melters. The agitation also prevents the formation of gas pockets, thus ensuring that hydrogen gas does not build up.

The research program for the project has been focused for nearly three years on testing and evaluating the effectiveness of PJMs. CFD models of all process vessels in the WTP have been developed either to confirm that the PJMs meet stringent mixing criteria, or, if performance is inadequate, to provide insight as to how the systems should be redesigned. State-of-the-art multiphase modeling techniques were used to prove out the basic design for fluidics mixing in the various process vessels. These models include the transient effects of solid-liquid mixing, such as accumulation, non-Newtonian yielding, air sparging, and heat transfer (Figure 2). By using CFD in this way, the project team was able to bypass extensive demonstration tests and keep pace with the plant's construction schedule.

FIGURE 2 (Left to right) Waste-processing tanks in a section of WTP. A plot taken from a CFD analysis of solids suspension in a vessel using PJMs showing settling during suction. Liftoff during drive. Source: Bechtel National, Inc.

DESIGN OF THE CHERNOBYL NEW SAFE CONFINEMENT

The Chernobyl new safe confinement (NSC) will shield the sarcophagus, or shelter, that was constructed soon after the nuclear accident in 1986 to contain the deadly radioactive materials in damaged Unit 4. Specifically, the NSC is designed to keep radioactive dust in and rain out and to facilitate the deconstruction of the sarcophagus and Unit 4. The NSC is intended to minimize occupational exposure for at least 100 years, with the expectation that improved storage or disposal methods will be available within that time.

The design team chose a movable, arch-shaped building made of large, preassembled pieces that could be constructed and then slid into place over the Unit 4 shelter (Figure 3). After considering eight initial configurations, the team selected one that met several of its key criteria and then optimized the configuration for chord depth and diameter and section shape. The layer of contaminated topsoil will be removed before construction begins to minimize radiation exposure to workers and schedule risk.

For rapid prototyping, the project team used state-of-the-art computer 3-D animation and virtual reality (VR) development software. The VR team was given a series of hand-drawn blueprints, annotated in Russian, along with ground-level and aerial photographs of the site and video footage of the area. Using the blueprints, they were able to create an accurate 3-D representation of the building exterior. Details of the confinement structure were provided by way of 2-D CAD drawings, which were used as a template to create a 3-D model. The heavily damaged interior was modeled primarily from photographs and video footage. The project team quickly developed 3-D simulations and generated large-scale animations by using distributed-rendering technology. The dynamic

FIGURE 3 Chernobyl NSC conceptual design shows the assembled arch positioned over the existing Unit 4 reactor. Source: Bechtel National, Inc.

view of the NSC construction and operations provided by the simulations greatly facilitated the project review.

SUMMARY

Because the cost of using simulation has decreased dramatically in recent years, it is now typically well within the cost and schedule constraints of many project budgets. Thus, engineering teams can now study the impact of environmental factors, including extreme conditions, from various perspectives. Prototyping the technical and conceptual aspects of projects on the computer early on can minimize many potential downstream risks.

ACKNOWLEDGMENTS

I would like to acknowledge the outstanding work of the engineers responsible for the projects discussed in this paper, including Brigette Rosendall, Kristian Debus, Carl Johnson, Chris Barringer, Martin Melin, and Feng Wen. They, along with their peers in the Advanced Simulation and Analysis Group at Bechtel, continuously strive to achieve success on challenging, schedule-driven assignments.

The Challenges of Landing on Mars

Tommaso P. Rivellini
Jet Propulsion Laboratory
California Institute of Technology
Pasadena, California

People have been fascinated with the idea of exploring Mars since the very beginning of the space age. Largely because of the belief that some form of life may have existed there at one time, surface exploration has been the ultimate ambition of this exploration. Unfortunately, engineers and scientists discovered early on that landing a spacecraft on the surface of Mars would be one of the most difficult and treacherous challenges of robotic space exploration.

Upon arrival at Mars, a spacecraft is traveling at velocities of 4 to 7 kilometers per second (km/s). For a lander to deliver its payload to the surface, 100 percent of this kinetic energy must be safely removed. Fortunately, Mars has an atmosphere substantial enough for the combination of a high-drag heat shield and a parachute to remove 99 percent and 0.98 percent respectively of the kinetic energy. Unfortunately, the Martian atmosphere is not substantial enough to bring a lander to a safe touchdown. This means that an additional landing system is necessary to remove the remaining kinetic energy.

On previous successful missions, the landing system consisted of two major elements, a propulsion subsystem to remove an additional 0.002 percent (~50 to 100 meters per second [m/s]) of the original kinetic energy and a dedicated touchdown system. The first-generation Mars landers used legs to accomplish touchdown. The second generation of touchdown systems used air bags to mitigate the last few meters per second of residual velocity. The National Aeronautics and Space Administration (NASA) is currently developing a third-generation landing system in an effort to reduce cost, mass, and risk while simultaneously improving performance as measured by payload fraction to the surface and the roughness of accessible terrain.

LEGGED LANDING SYSTEMS

The legs of the 1976 Viking mission lander represent the first-generation landing system technology (Pohlen et al., 1977). Basic landing-leg technology was developed for the lunar Surveyor and Apollo programs in the early 1960s. In conjunction with a variable-thrust liquid propulsion system and a closed-loop guidance and control system, legs represented an elegant solution to the touch-down problem. They are simple, reliable mechanisms that can be added to an integrated structure that houses the scientific and engineering subsystems for a typical surface mission (Figure 1).

The first challenge for a legged system is to enable the lander to touch down safely in regions with rocks. For this the legs must either be long enough to raise the belly of the lander above the rocks, or the belly of the lander must be made strong enough to withstand contact with the rocks. Neither solution is attractive. Either the lander becomes top heavy and incapable of landing on sloped terrain or a significant amount of structural reinforcement must be carried along for the remote chance that the lander will directly strike a rock. The decreased stability because of the high center of mass is exacerbated if a mission carries a large

FIGURE 1   First-generation landing system used on the Viking lander, which landed on Mars in 1976.

FIGURE 2 Lunokhod Soviet lunar rover leaving the legged lander that delivered it to the surface.

rover to the surface. Because of the rover's configurational requirements, it is typically placed on top of the lander. The Soviet Lunokhod lunar landers (Figure 2) are an excellent example of this type of configuration.

A second major challenge of the legged-landing architecture is ensuring safe engine cutoff. To prevent the guidance and control system from inadvertently destabilizing the lander during touchdown, contact sensors have been used to shut down the propulsion system at the moment of first contact. On sloped terrain, this causes the lander to free fall the remaining distance, which can significantly increase the total kinetic energy present at touchdown and, in turn, decrease landing stability and increase mission risk. Implementation and testing of fault protection for engine cutoff logic has been, and continues to be, a difficult problem.

The first in-flight problem associated with engine shut off occurred on the lunar Surveyor lander mission when the propulsion system failed to shut off at touchdown, resulting in a significant amount of postimpact hopping. Fortunately, the terrain was benign, and the problem was not catastrophic. The second in-flight problem occurred on the Mars 98 lander mission when the engines were inadvertently shut off prematurely because of a spurious contact signal generated by the landing gear during its initial deployment. This problem resulted in a catastrophic loss of the vehicle. As a result, the Apollo missions all reverted to a man in the loop to perform engine shut off.

A third major challenge with a legged landing system for missions with rovers is rover egress. Once the lander has come to rest on the surface, the rover must be brought to the surface. For legged landers, a ramped egress system is the most logical configuration. Because rovers are bidirectional, the most viable arrangement has been considered two ramps, one at the front and one at the rear of the lander. The Soviet Lunokhod missions landed in relatively benign terrain, and in all cases, both ramps were able to provide safe paths for the rover. In the Mars Pathfinder mission, one of the two ramps was not able to provide a safe egress path for the Sojourner rover, but the second ramp did provide safe egress. For vehicles designed to explore a larger fraction of the Martian surface and, therefore, land in more diverse terrain, combinations of slopes and rocks could conceivably obstruct or render useless the two primary egress paths.

AIR-BAG LANDING SYSTEMS

The second-generation landing system was developed for the Mars Pathfinder mission and subsequently improved upon for the Mars Exploration Rover (MER) missions (Figure 3). These second-generation systems have a combination of fixed-thrust solid rocket motors and air bags to perform the touchdown task. The solid rocket motors, which are ignited two to three seconds prior to impacting the surface, slow the lander down to a stop 10 meters above the surface, from an initial velocity of approximately 120 meters/second. The lander is then cut away from the over-slung rockets and free falls for the remaining distance.

The air-bag system, which was developed to reduce cost and increase landing robustness, is designed to provide omnidirectional protection of the payload by bouncing over rocks and other surface hazards. Because the system can also

FIGURE 3 Second-generation landing system used on the Mars Pathfinder and MER landers.

right itself from any orientation, the challenge of stability during landing has been completely eliminated. Because the lander comes to rest prior to righting itself, the challenge of rock strikes has been reduced to strikes associated with the righting maneuver, which are significantly more benign. The challenge of thrust termination, in this case cutting the lander away from the rockets, remains but has been decoupled from the problem of landing stability. The problems of rover egress were addressed systematically on the MER missions; a triple ramp-like system provided egress paths in any direction, 360 degrees around the lander.

Although the air-bag landing system has addressed some of the challenges and limitations of legged landers, it has also introduced some challenges of its own. Horizontal velocity control using solid rockets and air-bag testing were significant challenges for both the Mars Pathfinder and MER missions.

THE SKY-CRANE LANDING SYSTEM

As Mars surface explorations mature, roving is becoming more important in the proposed mission architectures. The MER missions demonstrated the value of a fully functional rover not reliant on the lander to complete its surface mission. In the 2009 Mars Science Laboratory (MSL) and other future missions, the rover's capabilities and longevity will be extended. Future missions are also being designed to access larger areas of the planet and, therefore, will require more robust landing systems that are tolerant to slope and rock combinations that were previously considered too hazardous to land or drive on. The third-generation landing system, the sky-crane landing system (SLS), currently being developed for the MSL mission, will directly address all of the major challenges presented by the first- and second-generation landing systems. It will also eliminate the problem of rover egress.

SLS eliminates the dedicated touchdown system and lands the fully deployed rover directly on the surface of Mars, wheels first. This is possible because the rover is no longer placed on top of the lander. In the SLS, the propulsion module is above the rover, so the rover can be lowered on a bridle, similar to the way a cargo helicopter delivers underslung payloads (Figure 4).

The landing sequence for future missions will be similar to the Viking mission, except for the last several seconds when the sky-crane maneuver is performed (Figure 5). After separating from the parachute, the SLS follows a Viking-lander-like propulsive descent profile in a one-body mode from 1,000 meters above the surface down to approximately 35 meters above the surface. During this time, a throttleable liquid-propulsion system coupled with an active guidance and control system controls the velocity and position of the vehicle. At 35 meters, the sky-crane landing maneuver is initiated, and the rover is separated from the propulsion module. The rover is lowered several meters as the entire system continues to descend. The two-body system then descends the final few meters to set the rover onto the surface and cut it away from the propulsion

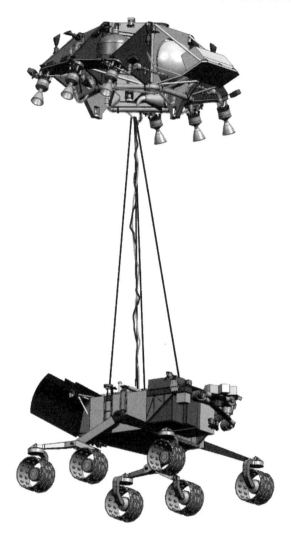

FIGURE 4 Sky-crane landing system shown with the rover already deployed.

module. The propulsion module then performs an autonomous fly-away maneuver and lands 500 to 1,000 meters away.

The central feature of the SLS architecture is that the propulsion hardware and terrain sensors are placed high above the rover during touchdown. As a result, their operation is uninterrupted during the entire landing sequence. One important result of this feature is that the velocity control of the whole system is improved, and, therefore, the rover touches down at lower velocity. Thus, there

FIGURE 5 Sky-crane landing sequence showing the three main phases and events. Nominal touchdown velocities are 0.75 m/s vertically and 0.0 m/s horizontally.

is no last-meter free fall associated with engine cutoff, and, because dust kick-up is minimal, the radar antennas can continue to operate even while the rover is being set down on the surface.

The lower impact velocity has two effects. First, the touchdown velocities can now be reliably brought down to the levels the rover has already been designed for so it can traverse the Martian surface. Second, the low velocity, coupled with the presence of bridles until the rover's full weight has been transferred to the surface, results in much more stability during landing.

Because the rover does not have to be protected from the impact energy at landing and because there is no need to augment stability at landing, there is no longer a need for a dedicated touchdown system. This, in turn, eliminates the need for a dedicated egress system. The SLS takes advantage of the fact that the rover's mobility system is inherently designed to interact with rough, sloping natural terrain. Rovers are designed to have high ground clearance, high static stability, reinforced belly pans, and passive terrain adaptability/conformability. These are all features of an ideal touchdown system.

TOUCHDOWN SENSING

Touchdown sensing can be done in several ways. The simplest and most robust way is to use a logic routine that monitors the commanded up-force generated by the guidance and control computer. The landing sequence is specifically designed to provide a constant descent velocity of approximately 0.75 m/s until touchdown has been declared. Prior to surface contact, the commanded up-force is equal to the mass of the rover plus the mass of the descent stage (which are roughly equal) times the gravity of Mars. During the touchdown event, the commanded up-force fluctuates depending on the specific geometry of the terrain.

Once the weight of the rover has been fully transferred to the surface of Mars, the commanded up-force takes on a new steady-state value equal to the mass of the descent stage times the gravity of Mars, approximately one-half of its pretouchdown magnitude. The system declares touchdown after the new lower commanded up-force has lasted for at least 1.5 seconds. This approach provides an unambiguous touchdown signature without the use of dedicated sensors.

The fly-away phase of the landing sequence is initiated when touchdown has been declared. During the fly-away phase, separation of the rover is accomplished by the pyrotechnic cutting of the bridle and umbilical lines connecting the rover and descent stage. The descent stage then uses its onboard computer to guide the propulsion module up and away from the rover and land it several hundred meters away.

CONCLUSION

As Mars explorers have learned the hard way, it's not typically the fall that kills you, it's the landing. Landing technology has matured significantly in the 40 years since NASA began exploring extraterrestrial surfaces. Each generation of landing technology has attempted to resolve the challenges posed by the previous generation. The SLS represents the latest stage in that evolution.

ACKNOWLEDGMENT

The research described in this paper was carried out at the Jet Propulsion Laboratory of the California Institute of Technology under a contract with the National Aeronautics and Space Administration.

REFERENCE

Pohlen, J., B. Maytum, I. Ramsey, and U.J. Blanchard. 1977. The Evolution of the Viking Landing Gear. JPL Technical Memorandum 33-777. Pasadena, Calif.: Jet Propulsion Laboratory.

Accessing the Lunar Poles for Human Exploration Missions

B. Kent Joosten
NASA Lyndon B. Johnson Space Center
Houston, Texas

The National Vision for Space Exploration calls for an American return to the Moon in preparation for the human exploration of Mars and other destinations. The surface environment of the Moon presents many challenges for human operations, but recent findings from robotic and Earth-based studies indicate that the lunar polar regions may offer advantages in terms of thermal conditions, availability of solar energy, and access to local resources. Although accessing these regions represents challenges in terms of orbital dynamics and propulsive performance, methods of accessing the polar regions are being actively investigated. Robotic missions are also being planned to gather environmental data on the poles and other areas of the Moon.

THE NATIONAL VISION FOR SPACE EXPLORATION

The National Vision for Space Exploration (NASA, 2004) laid out by President Bush in January 2004, calls for a return to the Moon in preparation for the human exploration of Mars and other destinations. According to this plan, the Moon will provide an operational environment for demonstrating human exploration technologies and capabilities within relatively safe reach of Earth. These capabilities included sustainable exploration techniques, such as the utilization of space resources, and human-scale exploration systems, such as power generation, surface mobility, and habitation and life support systems. In addition, lunar missions will pursue scientific investigations, such as gathering geological records of the early solar system.

ENVIRONMENT OF THE LUNAR POLES

Because the Moon's rotation rate is tidally locked to its revolution about the Earth, the Moon is a "slow rotator." The lunar diurnal period is 29.5 days. The long lunar night, combined with the absence of a lunar atmosphere, leads to large variations in surface temperature, from 117°C (243°F) to –170°C (-272°F) near the equator (Heiken et al., 1991). In addition to the severe cold during the lunar night, the lack of solar illumination for almost 15 days would mean an extended human exploration mission would have to rely upon nonphotovoltaic (i.e., nuclear) power sources.

During the Apollo Program, missions landed early in the lunar day and departed before the thermal conditions of lunar noon had set in. The polar regions of the Moon were essentially unexplored either by the Apollo Program or the Lunar Orbiter robotic missions of the 1960s. The polar regions were never considered as targets for the Apollo landings because of the constraints of orbital mechanics (i.e., high propulsive requirements and the absence of "free-return" abort capabilities). However, recent robotic and Earth-based observations of the Moon have revealed exciting new information about these previously unexplored regions.

Both the 1994 Clementine spacecraft and the Goldstone Solar System Radar have indicated the existence of permanently shadowed and nearly permanently lit areas in the rough surface topography in the vicinity of the lunar poles (Bussey et al., 1999; Margot et al., 1999). This is possible because of the very low inclination (only 1.5 degrees) of the lunar equator with respect to the ecliptic plane, which results in very shallow solar incidence. Illuminated areas of the poles are thought to have surface temperatures of –53°C (–63°F); the permanently shadowed regions would be only a few degrees above absolute zero (perhaps around –233°C or –387°F).

Both the shadowed and lit terrain, which appear to be in close proximity, are of interest for human lunar exploration. The nearly permanently illuminated areas offer moderate thermal conditions and potentially abundant solar power. The permanently shadowed regions have long been hypothesized to harbor "cold traps" that might have preserved volatiles, such as water ice deposited by millennia of cometary impacts. In fact, in 1998, the Lunar Prospector spacecraft detected trapped hydrogen, possibly in the form of water ice, in these regions (Feldman et al., 1998). If water were accessible in sufficient quantities, it would be a key contributor to sustained human presence on the Moon. Robotic lunar missions are planned to refine the topographic information and assess the true availability of resources.

ACCESSING THE LUNAR POLAR REGIONS

The Apollo missions did not attempt to use lunar polar regions as landing sites for several reasons. First, free-return trajectories, which allow the outbound trip to the Moon to be aborted and the spacecraft returned to Earth with little additional propulsion requirement, are not compatible with injection into lunar polar orbit. Polar orbits slowly drift into alignments that are not compatible with a return trajectory. Second, by choosing near-equatorial parking orbits and landing sites, crews could begin the return journey to Earth at nearly any moment with little propulsion performance penalty. Finally, there were strict limits on sun angles on the lunar sites during landing to ensure that the crew had good visibility of terrain relief.

Analyses are under way to meet these challenges. Future missions will address propulsion system failures by having additional levels of system redundancy and reliability rather than by free-return aborts. The orbital alignment issues can be addressed in several ways: (1) additional maneuvering capabilities can be built into the mission profile (with the associated propellant weight penalties); (2) an alternative to the Apollo "lunar orbit rendezvous" technique can be used, perhaps using Earth-Moon Lagrange points as a staging location; or (3) "safe havens" can be established on the lunar surface, thus negating the need for return to Earth at any given moment and making it possible to wait for more optimal return trajectories. We also expect that automated descent, landing, and hazard-avoidance technologies will obviate the need for strict lighting conditions.

CONCLUSION

Lunar polar locations may have environmental and resource advantages over equatorial sites and may enable future lunar mission to meet the goals laid out in the National Vision for Space Exploration. The challenges associated with human access to the polar regions are understood. With data from future robotic missions and advanced human mission capabilities and technologies, we should be able to address and overcome these challenges.

REFERENCES

Bussey, D.B.J, P.D. Spudis, and M.S. Robinson. 1999. Illumination conditions at the lunar south pole. Geophysical Research Letters 26(9): 1187–1190.

Feldman, W.C., S. Maurice, A.B. Binder, B.L. Barraclough, R.C. Elphic, and D.J. Lawrence. 1998. Fluxes of fast and epithermal neutrons from Lunar Prospector: evidence of water ice at the lunar poles. Science 281(5382): 1496–1500.

Heiken, G.H., D.T. Vaniman, and B.M. French. 1991. Lunar Sourcebook. Cambridge U.K.: Cambridge University Press.

Margot, J.L., D.B. Campbell, R.F. Jurgens, and M.A. Slade. 1999. Topography of the lunar poles from radar interferometry: a survey of cold trap locations. Science 284(5420): 1658–1660.

NASA (National Aeronautics and Space Administration). 2004. The Vision for Space Exploration. NP-2004-01-334-HQ. Washington, D.C.: NASA Headquarters. Available online at: < *http:// www.nasa.gov/pdf/55583main_vision_space_exploration2.pdf*>.

DESIGNER MATERIALS

Introduction

KRISTI S. ANSETH
*University of Colorado
Boulder, Colorado*

DIANN E. BREI
*University of Michigan
Ann Arbor, Michigan*

Over the past century, the materials field has evolved from the simple searching out, finding, and using of materials to the elegant, rational designing of highly functional materials, often assembled atom by atom, to impart desired and controlled properties. Materials designers have brought us smart materials, quantum wires and dots, diamond films and coatings, healing materials, and biomaterials.

Past scientific breakthroughs enabled us to visualize atoms inside materials and use quantum mechanics to explain how they interact to create bulk properties. Now we are heading down a path toward understanding not only how materials work, but also how they can be synthesized, manipulated, and processed in regulated ways on a molecular and atomistic level to create materials that can respond to changes in the environment in controlled and desirable ways. For example, biomaterials have evolved from off-the-shelf materials that served as passive, inert implants to sophisticated, rationally designed chemical structures that are biologically active, can control cell interactions and functions, and, in some cases, can actively promote healing. Smart materials have progressed from the natural crystals used in sonar during World War II to structural materials tailored for vibration, noise, and shape control of complex engineered systems, such as fighter jets and space optics.

Materials science involves applying a basic understanding to practical problems. Slowly but surely, materials scientists are becoming adept at using the tools of molecular genetics in the research laboratory. The only other place where materials are synthesized with such high fidelity, diversity, efficiency, and sophistication from a very limited set of building blocks is inside the cell! Materials scientists are beginning to exploit and mimic the evolved qualities of cells that enable their material production. Scientists can now readily control intracellular protein biosynthesis, which offers a general route to the engineering of macromolecular materials with precisely defined molecular weights, compositions, and sequencing. Materials scientists are also using living systems, such as the phage virus, as guides to the evolution of selected peptides that can recognize, bind, and grow electronic and magnetic building blocks.

Scientist and engineers are progressively bringing together natural process and innovative human ideas to design materials at all scales that will revolutionize how we engineer our world. At the nano-/microscale, materials can now be organized to share electrons between atoms to establish synergistic connections between phenomena. In multiferroric materials, for example, electric/magnetic interaction can be magnified by the transfer of energy between magnetic, electrical, and mechanical atoms. These materials will have real applications, such as futuristic radar systems for the battlefield and powerful cell phone antennas that will always be in contact—eliminating out-of-coverage areas! Another example is microscale "dimensioning," which can increase the surface area-to-volume ratio of components to provide rapid heat transfer. This could lead to the design of new materials, such as shape-memory alloy thin films, that can operate faster by two orders of magnitude, resulting in the highest known power densities for any solid-state actuation system. This technology could someday provide powerful micro-/mesoscale actuators for everything from new internal drug-delivery systems to systems that direct the nose cones of precision munitions.

Most current active materials can respond with one, or at most two, functions. But engineers envision extending the multifunctionality prevalent on the small scale in biological materials to larger scale engineering systems. Imagine the possibilities for structural materials with power, sensing, actuation, and processing/control capabilities! Imagine morphing aircraft profiles that can fly long distances as efficiently as birds, dive and attack if necessary, and land quietly. Imagine safety structures with crush zones during impact that can then heal themselves and return to operation. Imagine louvered or pore-based "smart skins" for temperature compensation and "comfort structures" that can eliminate noise and vibrations in cars or planes.

One day, perhaps soon, engineers will no longer be constrained by the properties of a material. Instead, they will define a material that fits the application requirements, thus opening the door to capabilities not yet thought of. The only limitation will be our imaginations.

Thin-Film Active Materials

GREG P. CARMAN
Department of Mechanical and Aerospace Engineering
University of California, Los Angeles
Los Angeles, California

This presentation describes recent progress and future trends associated with the development and potential uses of thin-film active materials. Active materials exhibit energy coupling, such as coupling between mechanical energy and electrical energy. Some widely recognized active materials are piezoelectric (electromechanical coupling), magnetostrictive (magneto-mechanical coupling), and shape-memory alloys (thermo-mechanical phase-transformation coupling). Many of the physical phenomena associated with bulk active materials are well known (cf., the work of the Curies in the 1800s). However, the same cannot be said about thin-film active materials, which have dimensions on the order of microns and represent a research area that is still in its scientific infancy.

Some physical-coupling phenomena that occur at small scales do not occur in bulk active materials. These unique phenomena include atomic coupling at the nanoscale regime and increased surface area-to-volume ratios, which could lead to the development of structures that were previously unimaginable. These advantages, as well as new active materials that are being discovered (e.g., ferro-magnetic shape-memory alloys), suggest that active thin-film material systems will be pervasive throughout our society in the upcoming decades (Murray et al., 2000). Applications for these unique phenomena range from powerful solid-state actuators to miniaturized sensors to clean power-generation systems. This presentation suggests how thin-film active materials will change our society.

In the Active Materials Laboratory at the University of California, Los Angeles (UCLA), research is being conducted on a wide range of active materials, including bulk materials and thin-film materials. One example is a thin-film shape-memory alloy composed of nickel (Ni) and titanium (Ti) (thermo-mechanical coupled). This class of shape-memory materials, discovered in bulk form in the 1960s, represents a benchmark material for shape-memory alloys. Although there are many other binary and ternary shape-memory systems, NiTi is the most often studied and the best characterized. The first publications on thin-film sputter-deposited NiTi appeared in the early 1990s (Busch and Johnson, 1990; Ikuta et al., 1990; Walker and Gabriel, 1990). Thus, research on thin-film shape-memory alloys is relatively new. Shape-memory materials undergo a solid-phase transformation from a low-temperature martensite phase to a higher temperature austenite phase. The phase-transformation temperature can be tailored by altering the composition of the material.

One unique property associated with NiTi is its ability to recover (actuate) a specific shape when heated through the austenite phase. For example, imagine a NiTi wire that is bent in the lower temperature martensite phase. When heated (e.g., using Joule heating), the wire quickly returns to its original straight configuration (i.e., shape memory). To illustrate this, Figure 1 shows a heating/cooling cycle for a thin-film NiTi system produced at UCLA. In Figure 1a, the film is at room temperature. From Figure 1a to 1c, the film is heated, and from Figure 1d to 1f, the film is cooled. This somewhat magical behavior is simply a phase transformation to a specific crystallographic arrangement resulting in "shape memory." This process, which converts thermal energy into mechanical energy, is used in a wide range of applications, including vascular stents (biocompatible), electrical connectors, satellite release bolts, and coffeepot thermostats.

Although shape-memory materials have unique attributes, the bandwidth (1 Hz) of bulk materials limits their applicability in many situations. The bandwidth limitation is due to relatively slow cooling processes related to bulk materials. However, thin-films have very large surface-to-volume ratios so their cooling times can be orders of magnitude higher.

Recently at UCLA, thin-film shape-memory alloy bandwidths higher than 100 Hz have been demonstrated (Shin et al., 2004); theoretical predictions approach the kHz regime. These large bandwidths, along with associated large stress and strain output, provide films with enormous values of power-per-unit mass (e.g., 40 kW/kg). The relatively large specific power (e.g., compared to <100 W/kg for other milli- or micro-motors) provides unique opportunities for small-scale applications (e.g., miniature motors and heart valves).

One pump design with dimensions in centimeters is used to articulate the nose cone of a small missile system (Shin et al., in press). Another application is a pumping motor that can move small amounts of fluids for various laboratory (e.g., laboratory-on-a-chip) and biomedical applications. For example, the pump can be used as an embeddable drug-delivery system. Coupled with appropriate sensors, this miniature (submillimeter) pump could someday replace a human

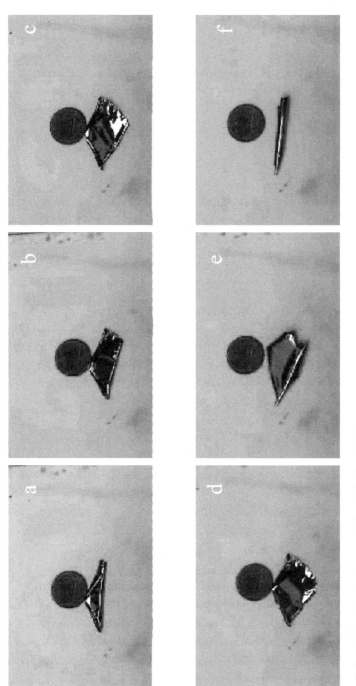

FIGURE 1 Heating/cooling cycle for a thin-film NiTi system.

pancreas in individuals suffering from diabetes. Recently, a cardiologist working with UCLA researchers proposed using this material for a percutaneous heart valve (Stepan et al., 2004). This revolutionary idea would mean that a heart valve could be replaced without major surgery. These are only a few of the future applications for thin-film shape-memory materials.

The potential for the general class of thin-film active materials is substantially larger than for thin-film shape-memory alloys alone. In the thin-film regime, atomic coupling can be used to improve the performance of active-material systems. For example, exchange coupling interaction, a phenomenon not present on the macroscale, can considerably decrease the magnetic field required to actuate magnetostrictive materials (Quandt and Ludwig, 1999). One advantage of magnetostrictive materials, noncontact actuation, can be generated because magnetic fields easily propagate through the air. For example, a permanent magnet easily rearranges iron particles on a plate without requiring direct physical contact with the particles.

A number of other submicron coupling phenomena could further improve thin-film active materials. For example, newly developed multiferroic systems that transform energies between multiple states (compared to classical two-state systems) are being studied (Zheng et al., 2004). Multiple-state energy transfer (e.g., mechanical to electrical power or solar to electrical power) is useful in constructing high-fidelity sensors, such as magnetometers that were once unimaginable and even novel, clean, power-generation systems that are far superior to existing systems (e.g., solar cells). These advancements will certainly change the way our world operates and will lead to many new scientific discoveries in the next century.

This presentation has described a few thin-film active materials and some of the concepts currently being pursued by academic and industrial researchers, such as the fabrication of, and applications for, new small-scale actuator and sensor systems. In the future, vast numbers of small-scale actuators may be massed together to achieve a common purpose, similar to the way ants interact during gathering operations. Another future application might be for small-scale systems to remove, or possibly prevent, blood clots, thereby preventing catastrophic strokes. In the more distant future, even-smaller-scale systems may interact and join together to form solid structures, similar to a fluid liquid that becomes a solid, morphing into different defined shapes. Material scientists are already designing materials at the atomic level, giving rise to the possibility that structures could be designed to interact and be built at the micron level. Therefore, the study of thin-film active materials, although still in its infancy, may revolutionize the future of our society.

REFERENCES

Busch, J.D., and A.D. Johnson. 1990. Prototype Micro-Valve Actuator. Pp. 40–41 in Proceedings of IEEE Micro Electro Mechanical Systems, An Investigation of Micro Structures, Sensors, Actuators, Machines and Robots. New York: IEEE.

Ikuta, K., H. Fujita, S. Arimoto, M. Ikeda, and S. Yamashita. 1990. Development of Micro Actuator Using Shape Memory Alloy Thin Film. Pp. 3–6 in Proceedings of the IEEE International Workshop on Intelligent Robots and Systems '90: Towards a New Frontier of Applications. New York: IEEE.

Murray S.J., M. Marioni, S. Allen, R.C. O'Handley, and T.A. Lograsso. 2000. 6% magnetic-field-induced strain by twin-boundary motion in ferromagnetic Ni-Mn-Ga. Applied Physics Letters 77(6): 886–888.

Shin, D., K.P. Mohanchandra, and G.P. Carman. 2004. High frequency actuation of thin film NiTi. Sensors and Actuators A: Physical 111(2-3): 166–171.

Shin, D., K.P. Mochanchandra, and G.P. Carman. In press. Development of hydraulic linear actuator using thin film SMA. Sensors and Actuators A: Physical.

Stepan, L., D. Levi, M. Fishbein, and G.P. Carman. 2004. A Thin Film Nitinol Heart Valve. Proceedings of the 2004 ASME International Mechanical Engineering Congress, Aerospace Division, Adaptive Structures and Material Systems Committee. New York: ASME.

Quandt, E., and A. Ludwig. 1999. Giant magnetostrictive multilayers. Journal of Applied Physics 85(8): 6232–6237.

Walker, J., K. Gabriel, and M. Mehregany. 1990. Thin-film processing of TiNi shape memory alloy. Sensors and Actuators A: Physical 21(1-3): 243–246.

Zheng, H., J. Wang, S.E. Lofland, Z. Ma, L. Mohaddes-Ardabili, T. Zhao, L. Salamanca-Riba, S.R. Shinde, S.B. Ogale, F. Bai, D. Viehland, Y. Jia, D.G. Schlom, M. Wuttig, A. Roytburd, and R. Ramesh. 2004. Multiferroic $BaTiO_3$-$CoFe_2O_4$ nanostructures. Science 303(5658): 661–663.

The Future of Engineering Materials: Multifunction for Performance-Tailored Structures

LESLIE A. MOMODA
HRL Laboratories, LLC
Malibu, California

In the future, new functional and reduced-scale materials that are currently in the forefront of technology will be hybridized into designer materials that can perform dramatic "tailorable" functions in large engineered systems. These performance-tailored structures will have the ability to change or adapt the performance or style of a structure on demand. Today, engineers can imagine designing adaptive flight profiles from morphing aircraft-wing structures; comfort-tailored performance, such as active structural vibration and noise suppression or temperature compensation, from louvered or pore-based "smart skins"; energy-efficient structures, such as tropical-plant-inspired solar structures; adaptive structures that can compensate for distortion or heal themselves; and structures reconfigured to satisfy style preferences. Imagine, for example, being able to commute to work in a stately professional car that can be reconfigured into a sportier car for the weekend.

As system-operating scenarios become more constrained by space and logistics limitations, the ability to adapt a structure's performance at will is becoming increasingly attractive. Currently multi-mission objectives are met with multiple structures (i.e., one car to drive to work and another one for the weekend). These solutions work if there is excess capacity in the system (e.g., a two-car garage), but as the number of mission objectives increases, the procurement, storage, and maintenance of a large number of structures become prohibitive. As a consequence, engineered subsystems that provide structural adaptability are under development in several programs, such as the Defense Advanced Research

Projects Agency (DARPA) Morphing Aircraft Structures Program (Wax et al., 2003), the General Motors Autonomy Concept (Burns et al., 2002), and many structural health-monitoring programs. All of these programs are designed to provide tailored performance in large multicomponent system structures.

Researchers are now thinking about how these same functionalities can be achieved in materials used to construct the system themselves, for example, a thin, smart-material skin that undergoes a radical but controlled change under mechanical strain; a coating that changes color on demand; a shell that reconfigures its shape to meet styling or performance criteria. These materials would enable the same system-level goals that are currently designed as sub-systems, but could be more easily integrated into larger engineered structures because they will be lighter, smaller, less difficult to interface, and easier to maintain than current subsystems. Driven by recent advances in biomaterials and nanotechnology, multifunctional materials are emerging as a new interdisciplinary field that promises to provide a new level of functionality, adaptability, and tailorability for future engineered systems.

MULTIFUNCTIONAL MATERIAL SYSTEMS

A multifunctional material is typically a composite or hybrid of several distinct material phases, in which each phase performs a different but necessary function, such as structure, transport, logic, or energy storage. Because each phase of the material performs an essential function, and because there is little or no parasitic weight or volume, multifunctional materials promise more weight-efficient, volume-efficient performance flexibility and potentially less maintenance than traditional multicomponent brass-board systems. The finer and more distributed the integration scale in the material, the faster and more autonomous the reaction times.

Multifunctionality in a material can be integrated on several dimensional scales with increasing interconnectivity between phases and engineering difficulty as the scale decreases. Matic (2003) has categorized these scales as different "types" (Figure 1). Type I material is comprised of phases in which one function is simply mounted, coated, or laminated to another, usually a structural component. Type II materials are comprised of distinct phases in which one function is embedded in another, usually a structural component. Type III materials are truly integrated; the phases are intermeshed, blurring the physical distinctions between them. The true promise of multifunctional materials for performance-tailored structures is most likely to be realized by Type III materials.

Multifunctional materials are designed for improved overall system performance. Thus, their performance metrics are inherently different from their single component phases, in which improvement of a single function, such as electrical conductivity, mechanical strain or force, or energy density, is maximized or minimized. A multifunctional material requires a new design methodology in

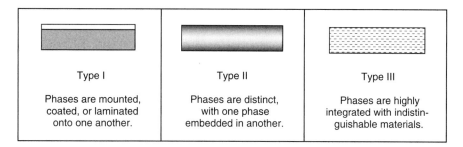

FIGURE 1 Schematic representations of the three types of multifunctional materials. Source: Matic, 2003. Reprinted with permission.

which system-level performance is emphasized over the optimization of individual functions. Optimizing system-level performance involves optimization methodologies that are not commonly used in materials science.

FRONTIERS IN MULTIFUNCTIONAL MATERIALS TECHNOLOGY

Materials Technology

The goal of multidisciplinary research on multifunctional materials, much of it under the auspices of the DARPA Synthetic Multifunctional Materials Program (Christodoulou and Venables, 2003), is to demonstrate weight and volumetric efficiencies and performance enhancements. Most research is focused on integrating two functions, usually a transport and structural function, and typically with low interconnectivity (e.g., Type I or Type II). Many of these studies rely heavily on inherently two-phase structural materials, such as fiber composites, laminates, foams, and other porous structures, as the matrix for multifunction. Even at this early stage, however, system-level benefits have been noted.

Structural batteries that reduce weight and complexity by integrating energy storage directly into the load-bearing structure have been developed by several teams using fibers, laminate, and nanotube construction. The energy density of the storage medium, such as a battery or supercapacitor, is reduced as the result of the incorporation of less conductive structural materials. However, the decrease in parasitic structure results in an overall weight savings and, therefore, improved energy density for the system as a whole.

The integration of actuation or sensing mechanisms into tailorable structural materials, which is essential for mechanical reconfigurability and structural morphing, is another area under investigation. Research using metallic foams or highly engineered mesostructured materials (dos Santos e Lucato et al., 2004) and elastomeric polymers (Pei et al., 2002) has shown tremendous promise for producing large structural motions along with the capacity to integrate actuation for feedback and control. Research conducted at HRL Laboratories, LLC, using hybrid, shape-memory materials (Figure 2), has demonstrated the feasibility of using integrated actuating tendons in a variably stiff matrix to produce low-energy flexure during reconfiguration but shape fixity for structural hold without energy when the motion is complete (W. Barvosa-Carter, personal communication). This concept has been extended to a hybrid laminate (Figure 3), which can produce a two-orders-of-magnitude change in stiffness, essentially from the stiffness of rubber to the hardness of steel, to accompany a shaping actuation force (G.P. McKnight, personal communication).

Self-healing composite structures are also under development (Chen et al., 2002). In these studies, a second phase, such as an adhesive or toughening agent, is added into the structure to ensure resealing on impact. Materials that can integrate other functions into structures, such as electromagnetism, thermal management, and optics capability, are also being investigated.

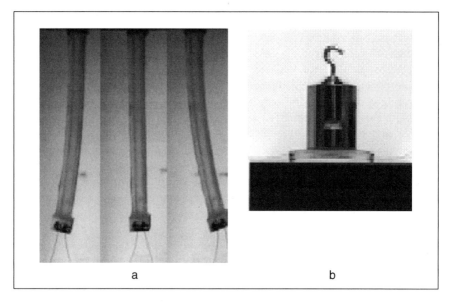

a   b

FIGURE 2 Hybrid shape-memory materials approach to reconfigurable structures. **a.** Shape-memory tendons actuate a variably stiff shape-memory polymer beam that can be flexed with low-input energy. **b.** Upon cooling, the beam becomes stiff, locking shape deformations in place to handle exterior loads (here 500 g).

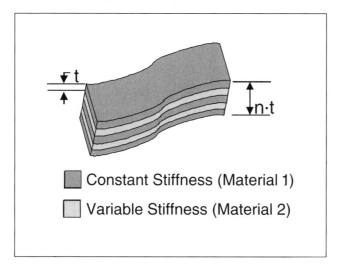

FIGURE 3 Variable stiffness laminate materials concept for producing large structural reconfigurations.

Optimization and Computational Design

Advances have been made in the development of optimization tools for designing integrated multifunctional materials. Sigmund and Torquato (1999) have done extensive work on topological optimization methods to determine the best morphological materials architectures to optimize performance from a highly integrated Type III material embedding of very dissimilar physical mechanisms. Many functional combinations, with as many as three phases, have been simulated. Although their work is purely theoretical, the results of their simulations have been validated by the similarity of the optimized topological solutions to microstructures and mesostructures found in biological systems. Macroscopic optimization tools for the design of less integrated Type I and II multifunctional materials have also been developed (Qidwai et al., 2002).

CHALLENGES AND PROSPECTS FOR THE FUTURE

Two-phase multifunctional systems show the promise of true materials integration. However, truly smart materials systems, analogous to biological systems, will require a combination of three or more functions, including logic, sensing, energy storage, structure, and actuation. Biological systems have perfected multifunction on a small scale. With the design of a priori multiple functions into a materials system, these concepts will be extended into large-scale structures. The complexities of these higher order systems will require a sophisticated understanding of how basic physical mechanisms can be manipulated to create new, potentially less singly optimal means of achieving function and multivariable optimization tools. For example, to increase the mechanical strength of energy storage systems, either the ionic conduction mechanism could be investigated for new logic capabilities or electrical conductivity mechanisms could be investigated to determine how they might influence mechanical strength.

As our understanding of materials on the nanoscale improves, we will be able to improve our control and increase the range of physical properties of materials even as we decrease the integration scale. Right now, while we are on the verge of understanding and harnessing physics on the nanoscale, we have a tremendous amount of work to do to learn how to fabricate large-scale materials from nanoscale elements. Although self-assembly and biological processing techniques look promising, they are not yet mature enough for the fabrication of multicomponent systems.

New system-level design methodology for materials will change not only the tools a materials scientist must know and understand, but also the role of the materials scientist in the system-design process. Typically, system designers choose from a toolbox of materials that have already been developed, and the materials scientist predesigns these materials to improve a single function. Often the materials scientist, who acts independently of the design team, is present only to characterize data or troubleshoot a problem after design. In the new paradigm, however, the materials scientist must be actively involved from the inception of the system design, providing a finely engineered material on the meso-, micro-, or nanoscale to meet the overall system goals. This will require that the materials scientist be completely familiar with system design tools and computational tools, from the system scale to the micro- or nanoscale. In the future, the design of a new car, airplane, or satellite will truly begin on the atomic scale.

REFERENCES

Burns, L.D., J.B. McCormick, and C.E. Borroni-Bird. 2002. Vehicles of change. Scientific American 287(4): 64–73.

Chen, X., M.A. Dam, K. Ono, A. Mal, H. Shen, S.R. Nutt, K. Sheran, and F. Wudl. 2002. A thermally re-mendable cross-linked polymeric material. Science 295(5560): 1698–1702.

Christodoulou, L., and J.D. Venables. 2003. Multifunctional material systems: the first generation. JOM: The Member Journal of The Minerals, Metals & Materials Society 55(11): 39–45.

dos Santos e Lucato, S.L., J. Wang, P. Maxwell, R.M. McMeeking, and A.G. Evans. 2004. Design and demonstration of a high authority shape morphing structure. International Journal of Solids and Structures 41(13): 3521–3543.

Matic, P. 2003. Overview of Multifunctional Materials. Pp. 61–69 in Smart Structures and Materials 2003, edited by D.C. Lagoudas. Active Materials: Behaviors and Mechanics. Proceedings of SPIE vol. 5053. Bellingham, Wash.: SPIE.

Pei, Q., R. Pelerine, S. Stanford, R. Kornbluh, M. Rosenthal, K. Meijer, and R. Full. 2002. Multifunctional Elastomer Rolls. Materials Research Society Symposium Proceedings 698: 165–170.

Qidwai, M.A., J.P. Thomas, and P. Matic. 2002. Design and Performance of Composite Multifunctional Structure-Battery Materials. Pp. 1–8 in Proceedings of the American Society for Composites Seventeenth Technical Conference, edited by C.T. Sun and H. Kim. Boca Raton, Fla: CRC Press.

Sigmund, O., and S. Torquato. 1999. Design of smart composite materials using topology optimization. Smart Materials and Structures 8(9): 365–379.

Wax, S.G., G.M. Fisher, and R.R. Sands. 2003. The past, present and future of DARPA's investment strategy in smart materials. JOM: The Member Journal of The Minerals, Metals & Materials Society 55(11): 17–23.

Biomimetic Strategies in Vascular Tissue Engineering

JENNIFER L. WEST
Departments of Chemical Engineering and Bioengineering
Rice University
Houston, Texas

INTRODUCTION

Cardiovascular disease, the leading cause of death in the United States, claims more lives each year than the next five leading causes of death combined (American Heart Association, 2003). Coronary heart disease caused more than one in every five American deaths in 2000 and required approximately 500,000 coronary artery bypass graft surgeries (CABGs) that year. Bypass grafting is also used in the treatment of aneurysmal disease or trauma.

At present, surgeons use autologous tissue and synthetic biomaterials as vascular grafts. Transplantation of autologous tissue has the best outcome in applications such as CABG because synthetic grafts, such as those made from expanded polytetrafluoroethylene or polyethylene terepthalate, fail in small-diameter applications (ID < 6 mm) due to the formation of blood clots and scar tissue. However, the supply of autologous tissue is often limited, either because of prior procedures or because of peripheral vascular disease. Recent advances in tissue engineering have raised hope that substitutes for blood vessels may one day be fabricated for small-diameter applications, such as CABG, where treatment options are often severely limited.

TISSUE ENGINEERING

Tissue engineering is the application of engineering principles to the design of tissue replacements, usually formed from cells and biomolecules. Tissue-

55

engineered products are already commercially available for skin and cartilage. Typically, an engineered tissue is formed by harvesting a small sample of the patient's cells, expanding them in culture, then seeding the cells onto a scaffold material. Scaffold materials, usually biodegradable synthetic polymers, are intended to define the size and shape of the new "tissue" and to provide mechanical support for the cells as they synthesize the new tissue. The cell-seeded scaffolds can then either be implanted into the patient, with tissue formation occurring in situ, or cultured further in vitro until their properties are more similar to those of normal tissue before implantation. This culture period is often carried out in a bioreactor to provide appropriate mechanical conditioning during tissue formation.

Most tissue-engineering strategies attempt to create small-caliber vascular grafts by closely mimicking the structure, function, and physiologic environment of native vessels. Normal arteries have three distinct tissue layers (Figure 1): the intima, the media, and the adventitia. The intima consists of a monolayer of endothelial cells that prevents platelet aggregation and regulates vessel permeability, vascular smooth muscle cell behavior, and homeostasis. In the medial layer, smooth muscle cells and elastin fibers aligned circumferentially provide most of the vessel's mechanical strength (Wight, 1996). The adventitial layer contains fibroblasts, connective tissue, the microvascular supply, and a neural

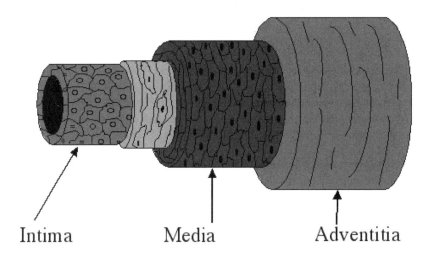

Intima   Media   Adventitia

FIGURE 1 The arterial wall is composed of three distinct layers: the intima, the media, and the adventitia. The intima, composed of endothelial cells, provides a nonthrombogenic surface. In the medial layer, smooth muscle cells and elastin fibers align circumferentially to provide mechanical integrity and contractility. The outer layer, the adventitia, is a supportive connective tissue.

network that regulates the vasotone of the blood vessel. The re-creation of some or all of the vessel layers and their properties may result in the development of a patent, functional vascular graft. In all likelihood, a successful graft will require an intima and a media.

As described above, the creation of a tissue-engineered vascular graft (TEVG) usually involves the harvest of desired cells, cell expansion in culture, cell seeding onto a scaffold, culture of the construct in an environment that induces tissue formation, and implantation of the construct back into the patient. Many options must be considered for each step in this process.

Efforts to create TEVGs remain in an early developmental stage, but several potential problems have already been identified. The patency of the graft is threatened by thrombosis, in all likelihood due to the retention of endothelial cells after implantation or alterations in endothelial cell function after culture in vitro. Also, the possibility of bursting after implantation in the physiological flow environment would have catastrophic consequences. Mechanical properties of TEVGs are generally lower than those of native arteries. Thus, a number of approaches are being tried to improve them. All of the strategies discussed above—cell source, genetic modification, scaffold materials, and culture conditions—are likely to impact the fabrication of an optimal, clinically useful TEVG.

Cell Sources for Vascular Tissue Engineering

The development of a functional TEVG is likely to require the construction of an intima and media composed of endothelial and smooth muscle cells. Limitations imposed by immunogenicity will probably require that autologous cells be used, so the majority of studies to date have used differentiated smooth muscle and endothelial cells isolated from harvested blood vessels. But problems with donor-site morbidity and the performance of these cell types in engineered tissues have led to the consideration of alternative cell sources. Recent advances in stem cell biology may lead to suitable progenitors that can be effectively differentiated into endothelial and smooth muscle cells for use in vascular tissue engineering.

Genetic Modification of Vascular Cells

Genetic engineering of vascular cells ex vivo may be an effective strategy for improving the properties of tissue-engineered grafts. The leading cause of failure in vascular grafts has been attributed to thrombosis, or the formation of blood clots. Seeding small-diameter vascular graft constructs with cells genetically engineered to secrete anti-thrombotic factors may improve graft patency rates.

A platelet aggregation inhibitor, nitric oxide (NO), for example, has shown promising results. Using an ex vivo approach, bovine smooth muscle cells liposomally-transfected with NO synthase III (NOS III) and GTP cyclohydrolase, which produces a cofactor essential for NOS activity, were grown as monolayers on plastic slides or biomaterials of interest and then placed in a parallel plate flow chamber. Whole blood was introduced into the flow chamber to assess platelet adherence to the cell monolayers. The number of platelets that adhered to the NOS-transduced smooth muscle cells was significantly lower than the number that adhered to mock-transduced smooth muscle cells and similar to the number that adhered to cultured endothelial cells (Scott-Burden et al., 1996).

The genetic modification of cells used to seed the TEVG may prove beneficial to improving the mechanical properties of TEVGs, which are related in large part to the composition and structure of the extracellular matrix (ECM). ECM cross-linking, which can result from the enzymatic activity of lysyl oxidase (LO) (Aeschlimann and Paulsson, 1991), may be a means of improving mechanical properties of the TEVG. LO, a copper-dependent amine oxidase, forms lysine-derived cross-links in connective tissue, particularly in collagen and elastin (Rucker et al., 1998). A gene therapy strategy has demonstrated that the mechanical properties of tissue-engineered collagen constructs are enhanced by using vascular smooth muscle cells transfected with LO (Elbjeirami et al., 2003). The elastic modulus and ultimate tensile strength of collagen gels seeded with LO-transfected smooth muscle cells nearly doubled compared to gels seeded with mock-transfected smooth muscle cells. The improved mechanical properties resulted from increased ECM cross-linking rather than increased amounts of ECM, changes in ECM composition, or increased cellularity. This strategy may ultimately lead to the enhancement of the mechanical characteristics of TEVGs and minimize the time required for in vitro culture prior to implantation.

Scaffolds for Vascular Tissue Engineering

Tissue engineers have had to choose either natural (e.g., collagen; Weinberg and Bell, 1986) or synthetic (e.g., polyglycolic acid; Niklason et al., 1999) polymer scaffolds. Each has advantages and disadvantages. Ideally, there would be specific cell-material interactions (like the interactions between cells and collagen), as well as control over material properties and the ease of processing that synthetic polymers offer.

Therefore, biomimetic derivatives of polyethylene glycol (PEG) are being studied as scaffolds for vascular tissue engineering. PEG-based materials are hydrophilic, biocompatible, and intrinsically resistant to protein adsorption and cell adhesion (Gombotz et al., 1991; Merrill and Salzman, 1983). Thus, PEG essentially provides a "blank slate," devoid of biological interactions, upon which the desired biofunctionality can be built. Because aqueous solutions of acrylated PEG can be rapidly photopolymerized in direct contact with cells and tissues (Hill-West et al., 1994; Sawhney et al., 1994), this is an easy method of cell seeding. Furthermore, PEG-based materials can be rendered bioactive by including proteolytically degradable peptides in the polymer backbone (West and Hubbell, 1999) and by grafting adhesion peptides (Hern and Hubbell, 1998) or growth factors (Mann et al., 2001a) into the hydrogel network during the photopolymerization process. Recently, PEG hydrogels that largely mimic the properties of collagen have been developed (Gobin and West, 2002; Mann et al., 2001b).

The elastin-derived peptide VAPG has been shown to be specific for smooth muscle cell adhesion, and PEG hydrogels modified with this adhesive peptide, rather than RGDS, support adhesion and the growth of vascular smooth muscle cells but not fibroblasts or platelets (Gobin and West., 2003). Moreover, bioactive molecules like TGF-β may be covalently incorporated into scaffolds to induce protein synthesis by vascular smooth muscle cells.

TGF-β has been reported to stimulate the expression of several matrix components, including elastin, collagen, fibronectin, and proteoglycans (Amento et al., 1991; Lawrence et al., 1994). TGF-β covalently immobilized to PEG-based hydrogels significantly increased collagen production of vascular smooth muscle cells seeded within these scaffold materials (Mann et al., 2001a). Mechanical testing of these engineered tissues also determined that the elastic modulus was higher in TGF-β-tethered PEG scaffolds than in PEG scaffolds without TGF-β, indicating that material properties for TEVGs may be improved using this technology. A cell-seeded graft formed from this biomimetic hydrogel scaffold is shown in Figure 2.

By using these types of bioactive materials, scientists may be able to capture the advantages of a natural scaffold, such as specific cell-material interactions and proteolytic remodeling in response to tissue formation, and take advantage of the benefits of a synthetic material, namely the ease of processing and the ability to manipulate mechanical properties.

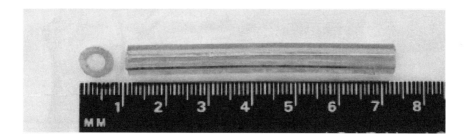

FIGURE 2 A PEG-based scaffold seeded with smooth muscle cells and endothelial cells ready for insertion into a bioreactor for in vitro culture of a TEVG (left, top view; right, side view). The cell-seeded scaffold is formed via photopolymerization, so the dimensions can be easily tailored for a given application, and cells are homogeneously seeded throughout the material. The scaffold is designed to degrade in response to cellular proteolytic activity during tissue formation.

Bioreactors for Mechanical Conditioning

In vivo, the pulsatile nature of blood flow puts radial pressure on the vessel wall, which subjects smooth muscle cells within the medial layer to cyclic strain. Thus, a great deal of research has focused on the response of smooth muscle cells to cyclic stretching. This research has demonstrated the importance of including cyclic strain during the fabrication of vascular tissue, particularly with respect to ECM synthesis and tissue organization. For example, smooth muscle cells seeded on purified elastin membranes and exposed to two days of cyclic stretching (10 percent beyond resting length) align perpendicular to the direction of applied strain; they also incorporate hydroxyproline into protein three to five times more rapidly than stationary controls, indicating increased collagen synthesis in response to strain (Leung et al., 1976).

Considering the profound effects of cyclic strain on the orientation of smooth muscle cells, ECM production, and tissue organization, preculture of vascular graft constructs in a pulsatile flow bioreactor system may help recreate the natural structure of native vessels and improve the mechanical properties of the construct. The mechanical stimuli from pulsatile flow could generate the cyclic strain necessary to alter ECM production, thereby creating a histologically organized, functional construct with satisfactory mechanical characteristics for implantation. Figure 3 is a schematic drawing of a typical pulsatile flow bioreactor system.

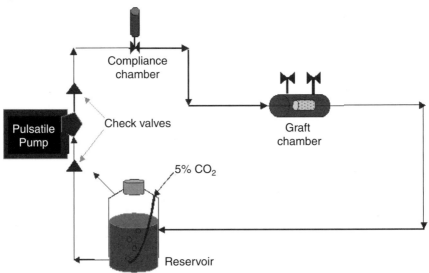

FIGURE 3 Diagram of a typical pulsatile-flow bioreactor for the culture of TEVGs. The pulse frequency and amplitude can be controlled via the pump, and the resultant strain environment can be controlled by altering the mechanical properties of the scaffold material.

To develop a blood vessel substitute, Niklason et al. (1999) cultured PGA constructs in a pulsatile blow bioreactor generating 165 beats per minute (bpm) and 5 percent radial strain. The pulse frequency of this system was chosen to mimic a fetal heart rate, which was believed to provide optimal conditions for the formation of new tissue. However, most investigations on mechanical conditioning have been conducted at 60 bpm, more representative of an adult heart rate, with promising outcomes. Therefore, the optimal bioreactor culture conditions for the development of a TEVG have not yet been determined. Nevertheless, this technology shows a great deal of promise for the production of a blood vessel substitute with the necessary mechanical and biochemical components.

CONCLUSION

Although a great deal of progress has been made on the creation of TEVGs, many challenges remain—particluarly preventing thrombosis and improving the mechanical properties of the graft. The development of a patent TEVG that grossly resembles native tissue now requires more than eight weeks of culture time. Thus, even with advances in the field, TEVGs are not likely to be used in emergency situations because of the time necessary for cell expansion, ECM production and organization, and the attainment of desired mechanical strength. Furthermore, unless advances in immune acceptance render the use of allogenic and xenogenic tissues feasible, autologous tissues will continue to be necessary to prevent an immunogenic response. TEVGs have not yet been subjected to clinical trials to determine their long-term efficacy. At that point, off-the-shelf availability and cost will be the biggest hurdles to the development of a feasible TEVG product.

Despite the many obstacles that must be overcome, the potential benefits of small-diameter TEVGs are exciting. In the near future, a non-thrombogenic TEVG with sufficient mechanical strength for clinical trials may be developed. Such a graft will have the minimum characteristics of biological tissue necessary to remain patent for a time comparable to current vein graft therapies.

As science and technology advance, TEVGs may evolve into complex blood vessel substitutes. They may become living grafts, capable of growing, remodeling, and responding to mechanical and biochemical stimuli in the surrounding environment. These blood vessel substitutes will closely resemble native vessels in their structure, composition, mechanical properties, and function. They will also have vasoactive properties, that is, they will be able to dilate and constrict in response to stimuli.

Close mimicry of native blood vessels may ultimately also be important in the engineering of other tissues that depend on vasculature to sustain function. As our understanding of the factors involved in cardiovascular development and function improves, we may one day develop TEVGs that will greatly improve the lives of people with vascular disease and other life-threatening conditions.

REFERENCES

Aeschlimann, D., and M. Paulsson. 1991. Cross-linking of laminin-nidogen complexes by tissue transglutaminase: a novel mechanism for basement membrane stabilization. Journal of Biological Chemistry 266(23): 15308–17.

Amento, E.P., N. Ehsani, H. Palmer, and P. Libby. 1991. Cytokines and growth factors positively and negatively regulate interstitial collagen gene expression in human vascular smooth muscle cells. Arteriosclerosis and Thrombosis 11(5): 1223–1230.

American Heart Association. 2003. Heart Disease and Stroke Statistics: 2003 Update. Available online at: *<http://www.americanheart.org/presenter.jhtml?identifier=4439>*.

Elbjeirami, W.M., E.O. Yonter, B.C. Starcher, and J.L. West. 2003. Enhancing mechanical properties of tissue engineered constructs via lysyl oxidase crosslinking activity. Journal of Biomedical Materials Research 66A: 513–521.

Gobin, A.S., and J.L. West. 2002. Cell migration through defined, synthetic extracellular matrix analogues. The Federation of American Societies for Experimental Biology Journal 16(7): 751–753.

Gobin, A.S., and J.L.West. 2003. Val-ala-pro-gly, an elastin-derived non-integrin ligand: smooth muscle cell adhesion and specificity. Journal of Biomedical Materials Research 67A(1): 255–259.

Gombotz, W.R., W. Guanghui, T.A. Horbett, and A.S. Hoffman. 1991. Protein adsorption to poly(ethylene oxide) surfaces. Journal of Biomedical Materials Research 25(12): 1547–1562.

Hern, D.L., and J.A. Hubbell. 1998. Incorporation of adhesion peptides into nonadhesive hydrogels useful for tissue resurfacing. Journal of Biomedical Materials Research 39(2): 266–276.

Hill-West, J.L., S.M. Chowdhury, A.S. Sawhney, C.P. Pathak, R.C. Dunn, and J.A. Hubbell. 1994. Prevention of postoperative adhesions in the rat by in situ photopolymerization of bioresorbable hydrogel barriers. Obstetrics and Gynecology 83: 59–64.

Lawrence, R., D.J. Hartmann, and G.E. Sonenshein. 1994. Transforming growth factor β1 stimulates type V collagen expression in bovine vascular smooth muscle cells. Journal of Biological Chemistry 269(13): 9603–9609.

Leung, D.Y.M., S. Glagov, and M.B. Mathews. 1976. Cyclic stretching stimulates synthesis of matrix components by arterial smooth muscle cells in vitro. Science 191(4226): 475–477.

Mann, B.K., R.H. Schmedlen, and J.L. West. 2001a. Tethered-TGF-β increases extracellular matrix production of vascular smooth muscle cells. Biomaterials 22 (5): 439–444.

Mann, B.K., A.S. Gobin, A.T. Tsai, R.H. Schmedlen, and J.L. West. 2001b. Smooth muscle cell growth in photopolymerized hydrogels with cell adhesive and proteolytically degradable domains: synthetic ECM analogs for tissue engineering. Biomaterials 22(22): 3045–3051.

Merrill, E.A., and E.W. Salzman. 1983. Polyethylene oxide as a biomaterial. American Society of Artificial Internal Organs Journal 6(2): 60–64.

Niklason, L.E., J. Gao, W.M. Abbott, K.K. Hirschi, S. Houser, R. Marini, and R. Langer. 1999. Functional arteries grown in vitro. Science 284(5413): 489–493.

Rucker, R.B., T. Kosonen, M.S. Clegg, A.E. Mitchell, B.R. Rucker, J.Y. Uriu-Hare, and C.L. Keen. 1998. Copper, lysyl oxidase and extracellular matrix protein cross-linking. The American Journal of Clinical Nutrition 67(5)(Suppl): 996S–1002S.

Sawhney, A.S., C.P. Pathak, and J.A. Hubbell. 1993. Bioerodible hydrogels based on photopolymerized poly(ethylene glycol)-*co*-poly(a-hydroxy acid) diacrylate macromers. Macromolecules 26: 581–587.

Scott-Burden, T., C.L. Tock, J.J. Schwarz, S.W. Casscells, and D.A. Engler. 1996. Genetically engineered smooth muscle cells as linings to improve the biocompatibility of cardiovascular prostheses. Circulation 94(Suppl II): II235–II238.

Weinberg, C.B., and E. Bell. 1986. A blood vessel model constructed from collagen and cultured vascular cells. Science 231(4736): 397–400.

West, J.L., and J.A. Hubbell. 1999. Polymeric biomaterials with degradation sites for proteases involved in cell migration. Macromolecules 32(1): 241–244.

Wight, T.N. 1996. Arterial Wall. Pp. 175–202 in Extracellular Matrix, Vol.1, edited by W.D. Comper. Amsterdam, Netherlands: Harwood Academic Publishers.

MULTISCALE MODELING

Introduction

GRANT S. HEFFELFINGER
Sandia National Laboratories
Albuquerque, New Mexico

DIMITRIOS MAROUDAS
University of Massachusetts
Amherst, Massachussetts

Multiscale modeling and simulation are used in scientific and engineering research in biology and the health sciences, materials and processing sciences, and earth and atmospheric sciences. This emerging integrated, computational approach can lead to better understanding, analysis, and quantitative predictions of the behavior of realistic, complex systems. Multiscale modeling and simulation links atomic-scale phenomena with macroscopic responses over time scales relevant to humans (from minutes to centuries) by establishing rigorous links between widely different theoretical formalisms and computational methods that capture a very broad range of space and time scales—from electronic, molecular, and mesoscopic or microstructural scales to continuum or macroscopic scales.

The core capabilities of multiscale modeling and simulation include computational quantum mechanics, statistical mechanics, and continuum mechanics combined with applied and computational mathematics, such as numerical analysis, nonlinear analysis of dynamic systems, optimization, and control theory. Some of these methods, which have been developed over many decades, are quite mature. However, the rigorous *coupling* of these methods to produce predictive models of phenomena that span multiple length and time scales remains a significant challenge. Examples range from linking molecular phenomena, such as DNA transcription and gene expression, with cellular, tissue, and organ response to connecting atomic-scale and grain-scale dynamics with the macroscopic response of engineering materials, such as metals, semiconductors, and polymers.

The ultimate goal of multiscale modeling and simulation is to produce a global framework for *system-level analyses* of processes and phenomena relevant to human scales that are governed by phenomena occurring at much finer length and time scales.

Achieving this goal will require advances in coupling various time and length scales, as well as in mathematics, computer science, and computational science. This session focuses on the state of the art in several aspects of multiscale modeling and simulation, including the coupling of modeling and simulation methods across time and length scales for specific applications in the science and processing of engineered materials (e.g., semiconductors, metals, and polymers) and in biology and health science. Presentations address recent advances in the theoretical, mathematical, and computer science underpinnings of multiscale modeling, as well as computational science challenges, such as tera- and peta-scale computing, advanced visualization, and enabling technologies.

Equation-Free Modeling for Complex Systems

Ioannis G. Kevrekidis
(with C.W. Gear and G. Hummer)
Department of Chemical Engineering, Program in Applied and Computational Mathematics, and Department of Mathematics Princeton University Princeton, New Jersey

ABSTRACT

In current modeling, the best available descriptions of a system are often given at a fine level (atomistic, stochastic, microscopic, individual-based) even though the questions asked and the tasks required by the modeler (prediction, parametric analysis, optimization and control) are at a much coarser, averaged, macroscopic level. Traditional modeling approaches first derive macroscopic evolution equations from the microscopic models and then bring an arsenal of mathematical and algorithmic tools to bear on these macroscopic descriptions. Over the last few years, and with several collaborators, we have developed and validated a mathematically inspired, computational enabling technology that allows the modeler to perform macroscopic tasks acting on the microscopic models directly. We call this the "equation-free" approach because it circumvents the step of obtaining accurate macroscopic descriptions. We argue that the basis of this approach is the design of (computational) experiments. Traditional continuum numerical algorithms can be viewed as protocols for experimental design (where "experiment" means a computational experiment set up and performed with a model at a different level of description). Ultimately, what makes the equation-free approach possible is the ability to initialize computational experiments at will. Short bursts of appropriately initialized computational experimentation— through matrix-free numerical analysis and systems theory tools like variance reduction and estimation— bridge microscopic simulation with macroscopic modeling. Remarkably, if there is enough control authority to initialize laboratory experiments "at will," this computational enabling technology can become a set of experimental protocols for the equation-free exploration of complex system dynamics.

A persistent feature of many complex systems is the emergence of macroscopic, coherent behavior from the interactions of microscopic "agents"—molecules, cells, individuals in a population—among themselves and with their environment. The implication is that *macroscopic rules*, a description of the system at a coarse-grained, high level, can somehow be deduced from *microscopic rules*, a description at a much finer level. For laminar Newtonian fluid mechanics, a successful coarse-grained description, the Navier-Stokes equations, was known on a phenomenological basis long before its approximate derivation from kinetic theory. Today we must frequently study systems for which the physics can be modeled at a microscopic, fine scale; but for whose macroscopic behavior explicit equations are practically impossible to derive. Hence, we look to the computer to explore the macroscopic behavior based on the microscopic description.

Macroscopic models of reaction and transport processes in our textbooks are in the form of conservation laws (e.g., species, mass, momentum, energy) closed through constitutive equations (e.g., reaction rates as a function of concentration, viscous stresses as functionals of velocity gradients). These models are written *directly* at the scale (alternatively, at the level of complexity) at which we are interested in modeling the system behavior. Because we observe the system at the level of concentrations or velocity fields, we sometimes forget that what really evolves during an experiment is distributions of colliding and reacting molecules.

We know from experience with particular classes of problems that it is possible to write predictive, deterministic laws for the behavior (predictive over relevant space/time scales that are useful in engineering practice) observed at the level of concentrations or velocity fields. Knowing the right level of observation at which we can be *practically predictive*, we attempt to write closed evolution equations for the system at this level. The closures may be based on experiment (e.g., through engineering correlations) or on mathematical modeling and approximation of what happens at more microscopic scales (e.g., the Chapman-Enskog expansion).

In many problems of current modeling practice, ranging from materials science to ecology and from engineering to computational chemistry, the physics are known at the microscopic/individual level, but the closures required to translate them to a high-level, coarse-grained, macroscopic description are not available. Sometimes we do not even know *at what level of observation* one can be practically predictive. Severe computational limitations arise in trying to bridge, through direct computer simulation, the enormous gap between the scale of the available description and the macroscopic, "system" scale at which the questions of interest are asked and the practical answers are required (see, e.g. Maroudas, 2000; Lu and Kaxiras, 2004). These computational limitations are a major stumbling block in current complex system modeling.

We will describe a computational approach for dealing with any complex, multiscale system whose collective, coarse-grained behavior is *simple* when we know in principle how to model the system at a very fine scale (e.g., through molecular dynamics). We assume that we do not know how to write good *simple* model equations at the right coarse-grained, macroscopic scale for the collective, coarse-grained behavior. We will argue that, in many cases, the derivation of macroscopic equations can be circumvented—that by using short bursts of appropriately initialized microscopic simulation, one can effectively solve the macroscopic equations *without ever writing them down*. Thus, a direct bridge can be built between microscopic simulation (e.g., kinetic Monte Carlo [kMC], agent-based modeling) and traditional continuum numerical analysis. It is possible to enable microscopic simulators to directly perform macroscopic, system-level tasks.

The main idea is to consider the microscopic, fine-scale simulator as a (computational) experiment that one can set up, initialize, and run at will. Based on the results of appropriately designed, initialized, and executed brief computational experiments, we can *estimate* the same information that a macroscopic model as we would *evaluate* from explicit formulas.

The heart of the approach can be conveyed through a simple example. Consider a single, autonomous ordinary differential equation,

$$\frac{dc}{dt} = f(c).$$

Think of it as a model for the dynamics of a reactant concentration in a stirred reactor. Equations like this embody "practical determinism": given a finite amount of information—the state at the present time, $c(t=0)$—we can predict the state at a future time. Consider how this is done on the computer using, for illustration, the simplest numerical integration scheme, forward Euler:

$$c_{n+1} \equiv c\left([n+1]\tau\right) = c_n + \tau f(c_n).$$

Starting with the initial condition, c_0, we go to the equation and *evaluate* $f(c_0)$, the time derivative, or slope of the trajectory $c(t)$. We then use this value to make a prediction of the state of the system at the next time step, c_1. We then repeat the process: go to the equation with c_1 to *evaluate* $f(c_1)$ and use the Euler scheme to predict c_2 and so on.

Forgetting for the moment accuracy and adaptive step-size selection, consider how the equation is used: given the state, we *evaluate* the time derivative; and then, using mathematics (in particular, Taylor series) and smoothness to create a local linear model of the process in time, we make a prediction of the state at the next time step. A numerical integration code will "ping" a subroutine with the current state as input and will obtain as output the time derivative at this state. The code will then process this value and use local Taylor series to make a prediction of the next state (the next value of c at which to call the subroutine evaluating the function f).

Three simple things are important to notice. First, the *task* at hand (numerical integration) does not need a closed formula for $f(c)$—it only needs $f(c)$ evaluated at a particular sequence of values c_n. Whether the subroutine evaluates $f(c)$ from a single-line formula, uses a table lookup, or solves a large subsidiary problem makes no difference from the point of view of the integration code. Second, the sequence of values c_n at which we need the time derivative evaluated is not known *a priori*. It is generated as the task progresses, *from processing the results of previous function evaluations* through the Euler formula. We know that protocols exist for designing experiments to accomplish tasks such as parameter estimation (Box et al., 1978). In the same spirit, we can think of the Euler method, and of explicit numerical integrators in general, as *protocols for specifying where to perform function evaluations* based on the task we want to accomplish (computation of a temporal trajectory). Last, the form of the protocol (the Euler method here) is based on mathematics, particularly on smoothness and Taylor series. The trajectory is locally approximated as a linear function of time; the coefficients of this function are obtained from the model using *function evaluations*.

Suppose now that we do not have the equation, but *we have the experiment itself*. We can fill up the stirred reactor with reactant at concentration c_0, run it for some time, and record the time series of $c(t)$. Using the results of a short run (over, say, one minute), we can now *estimate* the slope, dc/dt at $t=0$, and predict (using the Euler method) where the concentration will be in, say, 10 minutes. Now, instead of waiting nine minutes for the reactor to get there, we stop the experiment and immediately start a new one. We reinitialize the reactor *at the predicted concentration*, run for one more minute, and use forward Euler to predict what the concentration will be 20 minutes down the line. We are substituting short, appropriately initialized experiments, and *estimation* based on the experimental results, for the function evaluations that the subroutine with the closed form $f(c)$ would return. We are in effect doing forward Euler again, but the coefficients of the local linear model are obtained using experimentation "*on demand*" (Cybenko, 1996) rather than function evaluations of an *a priori* available model.

Now we complete the argument. Suppose that the inner layer is not a laboratory experiment but a *computational* experiment with a model at a different, much finer level of description—for the sake of the discussion, a lattice kMC model of the reaction. Instead of running the kMC model for long times and *observing* the evolution of the concentration, we can exploit the procedure described above, perform only short bursts of appropriately initialized microscopic simulation, and use the results to evolve the macroscopic behavior over longer time scales.

It is much easier to initialize *a code* at will—a computational experiment—than to initialize a new laboratory experiment. Many new issues arise, notably noise in the form of fluctuations, from the microscopic solver. The conceptual point, however, remains: even if we do not have the right macroscopic equation for the concentration, we can still *perform its numerical integration* without obtaining it in closed form. The skeleton of the wrapper (the integration algorithm) is the same one we would use if we had the macroscopic equation, but now function evaluations are substituted by short computational experiments with the microscopic simulator, whose results are appropriately processed for local macroscopic identification and estimation. If a large separation of time scales exists between microscopic dynamics (here, the time we need to run kMC to estimate dc/dt) and the macroscopic evolution of the concentration, this procedure may be significantly more economical than direct simulation.

Passing information between the microscopic and macroscopic scales at the beginning and the end of each computational experiment is vitally important. This is accomplished through a *lifting operator* (macro- to micro-) and a *restriction operator* (micro- to macro-) as discussed below (Theodoropoulos et al., 2000; Kevrekidis et al., 2003 and references therein). The proposed computational methodology consists of the following basic elements:

• Choose the statistics of interest for describing the long-term behavior of the system and an appropriate representation for them. For example, in a gas simulation at the particle level, the statistics would probably be density and momentum (zeroth and first moment of the particle distribution over velocities), and we might choose to discretize them in a computational domain via finite elements. We call this the macroscopic description, u. These choices suggest possible *restriction* operators, M, from the microscopic-level description U, to the macroscopic description: $u = MU$.

• Choose an appropriate *lifting* operator, μ, from the macroscopic description, u, to one or more consistent microscopic descriptions, U. For example, in a gas simulation using pressure etc. as the macroscopic-level variables, μ could make random particle assignments consistent with the macroscopic statistics, $\mu M = I$ (i.e., lifting from the macroscopic to the microscopic and then restricting [projecting] down again should have no effect, except roundoff).

• Start with a macroscopic condition (e.g., concentration profile) $u(t_0)$.

• Transform it through lifting to one or more fine, *consistent* microscopic realizations $U(t_0) = \mu u(t_0)$.

- Evolve this(these) realization(s) using the microscopic simulator for the desired short macroscopic time T, generating the value(s) $U(T)$;
- Obtain the restriction(s) $u(T)=MU(T)$ (and average over them).

This constitutes the *coarse time-stepper*, or *coarse time-T map*. If this map is accurate enough, it can be used as I described in a two-tier procedure to perform *coarse projective integration* (Gear and Kevrekidis, 2003; Gear, 2001; Gear et al., 2002) (see Figure 1a). Coarse projective integration, and also coarse bifurcation computations (see Figure 1b), have been used to accelerate lattice kMC simulations of catalytic surface reactions (Makeev et al., 2002a,b; Rico-Martinez et al., 2004), Brownian dynamics simulations of nematic liquid crystals (Siettos et al., 2003), and much more.

Time-stepper-based methods are, in effect, alternative ensembles for performing microscopic (molecular dynamics, kMC, Brownian dynamics) simulations. Innovative multiscale/multilevel techniques proposed over the last decade that can be integrated in an equation-free, time-stepper-based framework include the quasi-continuum methods of Phillips and coworkers (Phillips, 2001; Ortiz and Phillips, 1999) and the optimal prediction methods of Chorin and coworkers (Chorin et al., 1998, 2000) (see the discussion in Kevrekidis et al., 2003).

If one has good macroscopic equations, one should use them. But when these equations are not available in closed form, and such cases arise with increasing frequency in contemporary modeling, the equation-free computational enabling technology we have outlined here may hold the key to engineering *effectively simple* systems.

ACKNOWLEDGMENTS

This work was partially supported over the years by AFOSR, through an NSF/ITR grant, DARPA, and Princeton University. An extended version of this article appeared as a "Perspective" in the July 2004 issue of the *AIChE Journal* (with C. W. Gear and G. Hummer as coauthors).

REFERENCES

Box, G.E.P., W. Hunter, and J.S. Hunter. 1978. Statistics for Experimenters: An Introduction to Design, Data Analysis and Model Building. Indianapolis, Ind.: Wiley-Interscience.
Chorin, A., A. Kast, and R. Kupferman. 1998. Optimal prediction for underresolved dynamics. Proceedings of the National Academy of Sciences 95(8): 4094–4098.
Chorin, A., O. Hald, and R. Kupferman. 2000. Optimal prediction and the Mori-Zwanzig representation of irreversible processes. Proceedings of the National Academy of Sciences 97(7): 2968–2973.
Cybenko, G. 1996. Just in Time Learning and Estimation. Pp. 423-434 in Identification, Adaptation and Learning: The Science of Learning Models from Data, edited by S. Bittanti and G. Picci. NATO Advanced Studies Institute (ASI) Series F153. Berlin: Springer-Verlag.

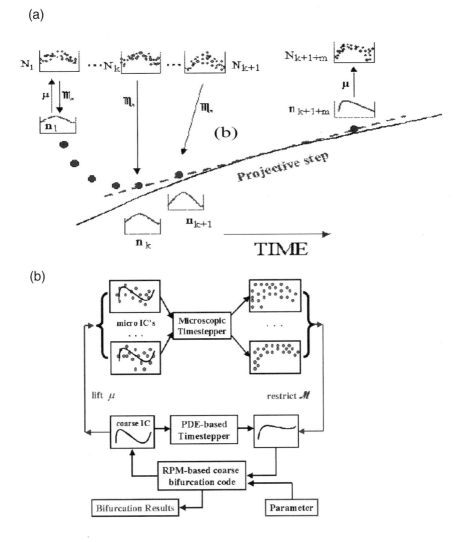

FIGURE 1 Schematic illustrations of (a) coarse projective integration and (b) coarse time-stepper-based bifurcation computations. Source: Gear et al., 2002. Reprinted with permission.

Gear, C.W. 2001. Projective Integration Methods for Distributions. NEC Research Institute Technical Report 2001-130. Princeton, N.J.: NEC Research Institute. Available online at: <*http:// www.neci.nj.nec.com/homepages/cwg/pdf.pdf*>.

Gear, C.W., I.G. Kevrekidis, and K. Theodoropoulos. 2002. Coarse integration/bifurcation analysis via microscopic simulators: micro-Galerkin methods. Computer and Chemical Engineering 26(7-8): 941–963.

Gear, C.W., and I.G. Kevrekidis. 2003. Projective methods for stiff differential equations: problems with gaps in their eigenvalue spectrum. Society for Industrial and Applied Mathematics (SIAM) Journal on Scientific Computing 24(4): 1091–1106.

Kevrekidis, I.G., C.W. Gear, J.M. Hyman, P.G. Kevrekidis, O. Runborg, and K. Theodoropoulos. 2003. Equation-free coarse-grained multiscale computation: enabling microscopic simulators to perform system-level tasks. Communications in Mathematical Sciences 1(4): 715–762.

Lu, G., and E. Kaxiras. 2004. An overview of multiscale simulations of materials. Available online at: <*http://arxiv.org/abs/cond-mat/0401073*>.

Makeev, A., D. Maroudas, and I.G. Kevrekidis. 2002a. Coarse stability and bifurcation analysis using stochastic simulators: kinetic Monte Carlo examples. Journal of Chemical Physics 116(23): 10083–10091.

Makeev, A.G., D. Maroudas, A.Z. Panagiotopoulos, and I.G. Kevrekidis. 2002b. Coarse bifurcation analysis of kinetic Monte Carlo simulations: a lattice gas model with lateral interactions. Journal of Chemical Physics 117(18): 8229–8240.

Maroudas, D. 2000. Multiscale modeling of hard materials: challenges and opportunities for chemical engineering. American Institute of Chemical Engineers Journal 46(5): 878–882.

Ortiz, M., and R. Phillips. 1999. Nanomechanics of defects in solids. Advances in Applied Mechanics 36: 1–79.

Phillips, R. 2001. Crystals, Defects and Microstructures. Cambridge, U.K.: Cambridge University Press.

Rico-Martinez, R., C.W. Gear, and I.G. Kevrekidis. 2004. Coarse projective KMC integration: forward/reverse initial and boundary value problems. Journal of Computational Physics 196(2): 474–489.

Siettos, C., M.D. Graham, and I.G. Kevrekidis. 2003. Coarse Brownian dynamics for nematic liquid crystals: bifurcation, projective integration and control via stochastic simulation. Journal of Chemical Physics 118(22): 10149–10156.

Theodoropoulos, K., Y.-H. Qian, and I.G. Kevrekidis. 2000. Coarse stability and bifurcation analysis using timesteppers: a reaction diffusion example. Proceedings of the National Academy of Sciences 97(18): 9840–9843.

Modeling the Stuff of the Material World: Do We Need All of the Atoms?

Rob Phillips
Division of Engineering and Applied Science
California Institute of Technology
Pasadena, California

The advent of computers ushered in a new way of doing science and engineering in which a host of complex problems ranging from weather prediction to the microstructural evolution of multiphase alloys to the DNA/protein interactions that mediate gene regulation could be explored explicitly using computer simulation. Indeed, some say that the physical sciences are now based on a triumvirate of experiment, theory, and simulation, with simulation complementing more traditional techniques for understanding problems involving many interacting degrees of freedom. One class of problems for which simulation is increasingly important is associated with the understanding and control of materials. When we speak of materials, we mean "stuff" as diverse as the materials of which man is made (soft, squishy stuff) and technologies (stuff with desirable properties, such as strength or conductivity) (Amato, 1997).

Clearly, the use of computation to understand and even design complex materials is one of the major challenges that will make it possible to replace the enlightened empiricism that gave rise to the great material ages (e.g., the Iron Age, Bronze Age, and silicon-based Information Age) with rational design. Similar roles are anticipated for simulation in many other fields as well. One of the flagship techniques for examining problems involving complex materials is molecular dynamics in which the microscopic trajectories of each and every atom are followed explicitly. Despite their promise, however, these simulations sometimes generate enormous quantities of information (i.e., terabyte data sets) without necessarily delivering the promised concomitant increase in understanding.

Terabyte data sets engendered by simulations represent a staggering quantity of information. A simple estimate reveals that the entire 10 floors worth of books in the Caltech Millikan Library corresponds roughly to a terabyte of information. More impressively, the genomes of many viruses have an information content that can be stored comfortably on a 256-megabyte memory stick alongside the genomes of even more complex organisms from bacteria to yeast. Indeed, even organisms as complex as humans have genomes that are much smaller than a terabyte. And yet our computers are overflowing with terabyte data sets, and worse yet, discussions of petabyte data sets are becoming routine. For example, a molecular dynamics calculation on a 100,000-atom system run for only 10 nanoseconds, woefully inadequate for accessing most materials processes, already generates a terabyte worth of data. Clearly, there is a mismatch between the quality of information generated in our simulations and the information present in genomes and libraries.

The question of how to build quantitative models of complex systems with many interacting degrees of freedom is not new. Indeed, one of the threads through the history of physics, the development of continuum theories, resulted in two compelling examples of this kind of theory—elasticity and hydrodynamics. These theories share the idea of smearing out the underlying discreteness of matter with continuum field variables. In addition, with both theories, material properties are captured in simple parameters, such as elastic moduli or viscosity, which reflect the underlying atomic-level interactions without specifically mentioning atoms.

One lesson of these examples is that "multiscale modeling" is neither the exclusive domain of computational model building nor a fundamentally new idea. Indeed, in the deepest sense, the sentiment that animates all efforts at model building, whether analytical or computational, is of finding a minimal but predictive description of the problem of interest.

One feature that makes problems like those described here especially prickly is that they often involve multiple scales in space or time or both. An intriguing response to the unbridled proliferation of simulated data has been a search for streamlined models in which there is variable resolution. Many of the most interesting problems currently being tackled in arenas ranging from molecular biology to atmospheric science are those in which structures or processes at one scale influence the physics at another scale. As a response to these challenges, modelers have begun to figure out how to construct models in which the microscopic physics is maintained only where needed. One benefit of these approaches is that they not only reduce the computational burden associated with simulations of complex systems, but they also provide a framework for figuring out which features of a given problem dictate the way the "stuff" of interest behaves. Several examples of this type of thinking are described below.

Before embarking on a discussion of case studies, it is worth discussing the metrics that might be used in deciding whether or not a particular coarse-grained model should be viewed as a success. From the most fundamental point of view, the job of theoretical models is to provide a predictive framework for tying together a range of different phenomena. For example, in the case of elasticity described above, there are vast numbers of seemingly unrelated problems (from flying buttresses to the mechanical response of ion channels) that may be brought under the same intellectual roof through reference to Hooke's law. With elasticity theory, we can predict how the cantilever of an atomic-force microscope will deflect when tugging on a protein tethered to a surface. In this sense, elasticity theory has to be viewed as an unqualified success in the coarse-grained modeling of materials and shows just how high the bar has been raised for multiscale models worthy of the name.

A CASE STUDY IN MULTISCALE MODELING

The Quasicontinuum Method

One of the computational responses to problems involving multiple scales is multiresolution models that attempt to capture several scales at the same time. There has been great progress along these lines in recent years from a number of different quarters, and presently we will consider one example, namely the quasicontinuum method that permits the treatment of defects in crystalline solids.[1] The main idea of the quasicontinuum method is to allow for atomic-level detail in regions where interesting physical processes, such as dislocation nucleation, dislocation intersections, and crack propagation are occurring, while exploiting a more coarse-grained description away from the key action. The motivation for the method is based on a recognition that when treating defects in solids there are both long-range elastic interactions between these defects and atomic-scale processes involving the arrangements of individual atoms. What makes these problems so difficult is that both the short-range and long-range effects can serve as equal partners in dictating material response.

The numerical engine that permits a response to problems of this type is finite elements that allow for nonuniform meshes and introduce geometric constraints on atomic positions through the presence of interpolation functions (so-called finite-element shape functions). Just as those of us who learned how to interpolate on logarithm or trigonometric tables remember, the key idea of the finite-element procedure is to characterize the geometric state of the system by keeping track of the positions of a few key atoms that serve as nodes on the finite-element mesh. The positions of all other atoms in the system can be found, if needed, by appealing to simple interpolation.

---

[1]There are many articles on the quasicontinuum method, but the interested reader is invited to consult *www.qcmethod.com*, which has an extensive list of papers dealing with this method.

To simulate material response, geometry is not enough. We not only have to know where the nodes are, but also what forces act on them. To that end, the quasicontinuum method posits that the forces on the nodes can be obtained by appealing to interatomic potentials that describe interactions between individual atoms. Using the interpolated atomic positions, a neighborhood of atoms around each node is constructed, and the energies and forces are then computed using standard atomistic techniques. This is an elegant prescription because it ensures that the material response is strictly determined by the underlying microscopic physics without any ad hoc material assumptions. Once the geometric mesh has been constructed and the forces on the nodes computed, the simulation itself can take place by either minimizing the energy with respect to nodal coordinates or by using $F = ma$ physics to compute the trajectories of the system over time.

For a concrete example, consider a crystalline solid subjected to external loading in the form of an indenter like the one shown in Figure 1. The quasicontinuum philosophy is to discretize the system in such a way that there is full atomic resolution where the action is (such as beneath the indenter) and a select, representative subset of atoms that serve as nodes of the finite-element regions where all-atom resolution is surrendered. For the particular case of two-dimensional dynamical nanoindentation considered here, the calculation involves a total of 5,000 nodes as opposed to the 10^7 atoms that would be needed in a full atomistic calculation. This point is driven home in an even more compelling fashion in the case of a fully three-dimensional calculation for which the full atomistic calculation would have implicated in excess of 10^{11} atoms (Knap and Ortiz, 2003).

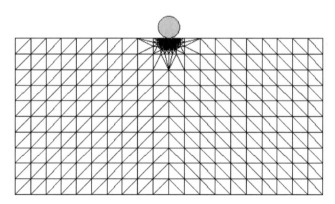

FIGURE 1 Schematic illustration of a multiresolution mesh used to describe nanoindentation of a crystalline solid. In the region just beneath the indenter, the mesh has full atomic resolution. In the "far fields," the mesh is much coarser.

The main point of this example is to reveal the kind of thinking now being used to address complex problems, such as material deformation. As exemplified by the quasicontinuum method, the underlying microscopic physics of bond stretching and bond breaking is treated explicitly where needed, and only approximately elsewhere.

THE PROBLEM OF LIVING MATERIALS

Understanding the workings of living materials presents even more compelling multiscale challenges than those encountered in the traditional materials setting. Indeed, we are now realizing that almost no individual macromolecule in the living world acts alone. Rather, the cell can be viewed as "a collection of protein machines," assemblies of individual macromolecules (Alberts, 1998). One of the pressing challenges to have arisen from the stunning successes of structural biology is the study of assemblies, such as viruses and the many *"SOMES"* (i.e., macromolecular assemblies that work in concert to maintain the mass and energy budget of the cell), such as nucleosomes, ribosomes, proteosomes, and assemblies that mediate gene expression, such as replisomes, spliceosomes, and so on.

Models of the function of assemblies such as *"SOMES,"* will remain out of reach of traditional atomic-level techniques for the foreseeable future. Consider the process of translation mediated by the ribosome. Even if we very generously assume that a new amino acid is added only once every millisecond, a molecular dynamics simulation of translation would have to be run for 10^{12} time steps for the addition of even a single amino acid to the nascent polypeptide. The number of atoms (including the surrounding water) engaged in this process is well in excess of 100,000, implying a whopping 10^{17} positions corresponding to all of the atoms during the entire molecular dynamics trajectory.

Similar estimates can be made for the workings of many other macromolecular assemblies that mediate the processes of a cell. All of these estimates lead to the same general conclusion—that even as we continue to pursue atomic-level calculations, we must redouble our efforts to understand the workings of *"SOMES"* from a coarse-grained perspective.

So the hunt is on to find methods of modeling processes of biological relevance involving assemblies of diverse molecular actors, such as proteins, lipids, and DNA, without having to pay the price in excessive data of all-atom simulation. One way to guarantee a rich interplay between experiment and models is through the choice of case studies that are well developed from the standpoint of molecular biology and for which we have compelling quantitative data.

One example of great importance is the *lac* operon, which has served as the "hydrogen atom" of gene regulation. This gene regulatory network, which controls the digestion of the sugar lactose in bacteria, has been the cornerstone of the development of our modern picture of gene regulation. An intriguing history of this episode in the history of molecular biology can be found in the books of Judson (1996) and Echols (2001).

The basic idea is that only when a bacterium is deprived of glucose and has a supply of lactose does the bacterium synthesize the enzymes needed to digest lactose. The "decisions" made by the bacterium are mediated by molecules, such as *lac* repressor, a protein that sits on the DNA and prevents the genes responsible for lactose digestion from being expressed. *Lac* repressor binds to several sites in the vicinity of the promoter for the genes responsible for lactose digestion and prohibits expression of those genes while simultaneously creating a loop of DNA between the two repressor binding sites.

From a modeling perspective, a minimal description of this system involves the DNA molecule itself, RNA polymerase, *lac* repressor, and an activator molecule called CAP. The kinds of questions that are of interest from a quantitative modeling perspective include the extent of gene expression as a function of the number of copies of each molecular actor in this drama, as well as the distance between the DNA binding sites for *lac* repressor and other features.

One recent multiscale attempt to simulate the interaction between DNA and *lac* repressor uses a mixed atomistic/continuum scheme, in which the *lac* repressor and the surrounding complement of water molecules are treated in full atomistic detail while the looped DNA region is treated using elasticity theory (Villa et al., 2004). The advantage of this approach is that it permits the DNA to present an appropriate boundary condition to the *lac* repressor simulation without having to do a full atomistic simulation of both DNA and the protein. Figure 2 shows an example of the simulation box and the elastic rod treatment of DNA. The key point of this example is not to illustrate what can be learned about the *lac* operon using mixed atomistic-continuum methods, but to illustrate how multiscale methods have begun to take root in the biological setting, just as they have in the conventional materials setting.

A second scheme, even more coarse-grained than the multiscale simulations of the *lac* repressor, is a statistical mechanics treatment of molecular decision makers, such as the repressor and its activator counterpart CAP. The relevant point is that all of the atomic-level specificity is captured by simple binding energies that reflect the affinity of these molecules for DNA and for each other.

This statistical mechanics perspective is a natural quantitative counterpart to the cartoons describing gene regulation used in classic texts of molecular biology. As shown in Figure 3, these cartoons depict various states of occupancy of the DNA in the neighborhood of the site where RNA polymerase binds. The statistical mechanics perspective adds the ability to reckon explicitly the statistical weights of each distinct state of occupancy of the DNA. From these statistical weights, a quantitative prediction can be made of the probability that a given gene will be expressed as a function of the number of molecules of each species.

Ultimately, one of the primary ways of judging a model must be by its ability to make predictions about as-yet undone experiments. The outcome of the

FIGURE 2 Illustration of a mixed atomistic/continuum description of the interaction of *lac* repressor protein with DNA. The lac repressor molecule is shown in dark gray, the part of the DNA treated explicitly is shown in medium gray, and the part of the DNA treated via continuum mechanics is shown as a ribbon. Water molecules surrounding the protein are shown in light gray. Source: Courtesy of Klaus Schulten and Elizabeth Villa.

so-called "thermodynamic models" (Ackers et al., 1982) described above is a predictive framework that characterizes the extent to which genes are expressed as a function of concentrations of the relevant decision-maker molecules (i.e., the transcription factors), the distance between the looping sites on the DNA, etc. Paradoxically, as a result of the great successes of structural biologists in determining the atomic-level structures of important complexes, such as DNA and its binding partners, we are now faced with the challenge of eliminating molecular details in models of their function.

SUMMARY

The critical question for building models of the material world is the extent to which we can suppress an atom-by-atom description of the function of materials. As emergence of "multiscale modeling" reveals, even with increasing

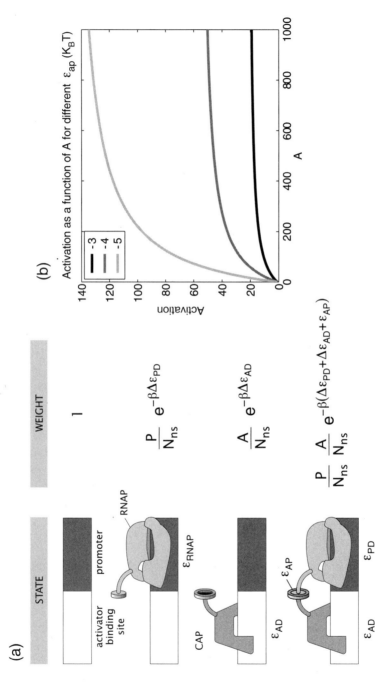

FIGURE 3 a. Schematic illustration showing the relation between cartoon models of various states of the genetic network and their corresponding weights in the statistical mechanics framework. **b.** Graph showing activation as a function of the number of activator molecules. The three curves correspond to different strengths for the interaction between the RNA polymerase and the activator.

computational power, a host of important problems remain out of reach of strictly brute-force approaches. In the analysis of the material world, whether of the complex, rigid metallic structures used to construct cities or the soft, squishy materials that make up the organisms that populate them, the key action often takes place at the level of individual atoms—whether of a bond breaking at a crack tip or the active site of an enzyme. Many of the atoms that are part of these processes are interlopers, however, that seem to be little more than passive observers that provide boundary conditions for the atoms actively involved in the process of interest.

From a model-building perspective, the goal is to "make things as simple as possible, but no simpler" (i.e., to eliminate as many molecular details as possible). In my opinion, this is one of the key design criteria for multiscale models. Although multiscale computational models are receiving most of the effort and attention right now, I believe the hunt should continue for analytic models that can capture the key features of complex materials and lead to the kind of insight that can be discussed at a blackboard.

ACKNOWLEDGMENTS

I am grateful to Jané Kondev, Michael Ortiz, Ellad Tadmor, Ron Miller, Vijay Shenoy, Vivek Shenoy, David Rodney, Klaus Schulten, Darren Segall, and Laurent Dupuy for collaboration and conversation. All of them played a key role in the development of my understanding of multiscale modeling. Farid Abraham first illustrated the meaning of a terabyte to me by comparing it to the informational content of a library.

REFERENCES

Ackers, G.K., A.D. Johnson, and M.A. Shea. 1982. Quantitative model for gene regulation by 1 phage repressor. Proceedings of the National Academy of Sciences 79(4): 1129–1133.

Alberts, B. 1998. The cell as a collection of protein machines: preparing the next generation of molecular biologists. Cell 92(3): 291–294.

Amato, I. 1997. Stuff: The Materials the World Is Made Of. New York: Basic Books.

Echols, H. 2001. Operators and Promoters. Berkeley: University of California Press.

Judson, H.F. 1996. The Eighth Day of Creation. Plainview, N.Y.: Cold Spring Harbor Laboratory Press.

Knap, J., and M. Ortiz. 2003. Effect of indenter-radius size on Au(001) nanoindentation. Physical Review Letter 90(22). Article No. 226102.

Villa, E., A. Balaeff, L. Mahadevan, and K. Schulten. 2004. Multi-scale method for simulating protein-DNA complexes. Multiscale Modeling and Simulation 2(4): 527–553.

Balancing Scales in Biological Models

ADAM PAUL ARKIN
Howard Hughes Medical Institute
Department of Bioengineering, University of California, Berkeley
Physical Biosciences Division,
E.O. Lawrence Berkeley National Laboratory
Berkeley, California

I cannot doubt but that these things, which now seem to us so mysterious, will be no mysteries at all; that the scales will fall from our eyes; that we shall learn to look on things in a different way—when that which is now a difficulty will be the only commonsense and intelligible way of looking at the subject

Lord Kelvin
Presidential Address to the Institution of
Electrical Engineers, 1889

Almost all of biology is about evolution. On short time scales, biologists study the kinetics of single enzymes and the temporal evolution of a product from a substrate. On longer time scales, they study the evolution of cellular behaviors that result from the integrated dynamics of the network of nonlinear, and possibly stochastic, chemical reactions that link genome to physiology and respond to environmental signals. On even longer time scales, they study how cells develop—changing their overall functions in response to the genetic program and organizing groups of cells into functional units, such as spore stalks, organs, and sometimes human beings. Scaling up even further, organisms operate as populations that spread beyond their niches in the world and compete with other organisms for resources to survive. This leads to the final scale, the true genetic evolutionary time scale, in which life emerges from the proverbial primordial soup and, on a time scale that defies intuition or complete conception, rewrites itself over and over with mistakes enough that new forms are born and old forms die on the shorter time scales just described.

To reconstruct the tree of life from the first bacterium, currently dated about 3.5 billion years ago, to modern metazoans like you and me, we peer as "through a glass, darkly" at the striking similarities among phylogenies of fossils and genetic sequences and infer the historical dynamics that created them and us.

From fossil evidence to functional genomics, the purpose of biological models has always been to make sense of these complex data and to follow the implications of conceptual theories.

Models generally play three roles in biology: to demonstrate or explain a physical effect or elucidate a principle (such as the Mendelian sorting of traits via the gene theory or the Hodgkins-Huxley model of signal propagation in neurons); to demonstrate consistency, as when a number of assertions in the form of biochemical reactions in a cell are combined into an integrated model of a process, such as bacterial chemotaxis, to show that these reactions are sufficient to explain a cellular behavior, such as exact adaptation to a step of chemoattractant; and to explore teleology, that is, to demonstrate why something is the way it is. For example, models may be used to explain why an integral feedback is necessary for small, free-swimming cells, such as flagellated bacteria, to sense and follow a chemical gradient.

In this talk, I discuss biological models at each time scale and how they are being linked through improved measurement technology and genetic sequencing. I then demonstrate the theoretical and computational challenges presented by all of these inherently multiscale models, such as time-scale separations, high dimensionality, and differing levels of detail about different parts of the system. I also discuss the effects of choosing different physical pictures—macroscopic or mesoscopic kinetics—on the computational feasibility of asking certain questions of the model and on predicting behavior.

I also discuss the particular challenges in the analysis (rather than simulation) of these systems, comparisons of models to data, and the visualizations of the results. Examples in this presentation include a model of the control and evolution of stress response in *Bacillus subtilis* and models of spatial signaling in immune-cell chemotaxis.

Small-Scale Processes and Large-Scale Simulations of the Climate System

Bjorn B. Stevens
Department of Atmospheric and Oceanic Sciences
University of California, Los Angeles
Los Angeles, California

Simulations of weather and climate have long posed challenges for computational science. Atmospheric and oceanic circulations operate on spatial scales ranging from micrometers or smaller to planetary scales and temporal scales ranging from microseconds to millennia and beyond. The representation of this range of scales is far beyond the capacity of any envisioned computational platform. For now and the foreseeable future, simulations of atmospheric and oceanic circulations will require a massive truncation of scale and hence a loss of information. Because the circulations of interest are turbulent, truncation is not a trivial matter. To prevent the truncation of information at some scale from entailing a loss of predictability at the remaining scales, procedures must be developed for representing the effects of the truncated, or unresolved, scales on those that remain.

The spectrum of energy in a system often serves as a guide to choosing which scales to keep and which to discard. As we know from common experience, variability is most pronounced on the largest scales (i.e., seasonal differences in weather are larger than day-to-day variations, and the weather varies more from continent to continent than it does from one side of town to another). Consequently, simulations of the climate system invariably begin by explicitly representing the largest spatial scales and working their way down the spectrum as computational resources permit. The high spatio-temporal correlation among atmospheric processes—small scales tend to be fast, and large scales tend to be slow—means that truncation also has a temporal projection.

Currently available computational models of the global atmosphere and ocean are typically restricted to a representation of their respective fluids by numerical meshes capable of sampling spatial scales on the order of 100 to 200 kilometers on the horizontal plane and perhaps 100 to 1,000 meters on the vertical plane. The very largest computers in existence can produce calculations on horizontal and vertical meshes with linear dimensions more refined by a factor of 10. But even these calculations leave an enormous range of scales unresolved.

Thus, one of the central questions in our field, and the focus of this paper, is whether the net effect of the smaller-faster scales on the larger-slower scales can be represented as a function of the larger-slower scales that are explicitly tracked in a simulation. Although this question is posed in purely practical terms, it has an esthetic quality in that any such representation can be thought of as a formalism of our understanding.

Atmospheric and oceanic scientists often use the word *parameterization* to describe this formalism. In our jargon, the goal is to *parameterize* the collective effects of small-scale processes on large-scale processes. Because small-scale processes are sundry, the parameterization problem is multifaceted. Typically, small-scale processes are broken down into distinct classes of problems—clouds, radiative transfer, hydrometeor interactions, surface interactions, small-scale turbulence, chemistry, and so on—processes that can be thought of as the atoms. Although one may be interested only in the net effect of all of these processes, atomization facilitates idealization and subsequent study.

An artifact of this kind of decomposition is that it raises the question of how, and on what scale, individual processes (atoms) (e.g., clouds, radiation, chemistry, etc.) interact, and hence the extent to which parameterizations must be coupled to one another, and not just to larger-scale processes. Thermodynamic analogies are useful to a point; for instance, diffusion parameterizes molecular transport in fluids. However, any attempt to develop a kinetic theory capable of aggregating many small-scale processes is impeded by our lack of understanding of what exactly constitutes the atoms and the rules that govern their behavior.

A conspicuous example of a parameterization in the atmosphere is for fluxes of heat, momentum, and matter from an interface (Garratt, 1992). Simply stated, the question is: Given the state of the large-scale flow above an interface, and a gross characterization of an interface, for instance a measure of its roughness, its temperature, etc., what are the fluxes of momentum, matter, and enthalpy from the interface? Physically these fluxes are carried by correlated fluctuations in the velocity and temperature fields—eddies—whose sizes scale with their distance from the interface. At the interface, small roughness elements (capillary waves on the ocean; rocks, sand, bushes, cars on land) disturb the flow, leading to small-scale pressure gradients around obstacles that accelerate the flow and generate eddies that in turn transport enthalpy and matter, which has diffused from the surface roughness elements into the fluid, into the interior of the fluid.

In almost any practical application, the net effect of these eddies must be represented in some fashion to provide meaningful boundary conditions for the larger-scale flow. Attempting to aggregate the fundamental solutions of equations for flow around an ensemble of obstacles has proven fruitless. Instead, a so-called similarity approach (Barenblatt, 1996) has been developed. A key aspect of the similarity approach is simplifying the problem to a point where it becomes empirically tractable and then hoping that the answers so derived are relevant to less idealized situations.

For the surface flux problem, the similarity approach usually consists of first considering flow over a uniformly rough wall in the absence of temperature differences; the essence of the flow can only conceivably be retained if only two variables are considered, namely the distance, z, from the surface and a velocity scale that measures the momentum flux, e.g.,

$$u_*^2 \equiv \overline{w'u'}. \tag{1}$$

Here $\{u, w\}$ denotes the horizontal and vertical components of the velocity field, and primes indicate deviations from a large-scale average denoted by an overbar. To the extent that u_* and z are the only relevant parameters, it is possible to argue on purely dimensional grounds that

$$\frac{d\bar{u}}{dz} = \frac{u_*}{z}\alpha \tag{2}$$

where α is a dimensionless constant.

The scaling law in Eq. (2) derives its simplicity, and hence its empirical tractability, from the neglect of a variety of other, potentially important, parameters. For instance, the formulation implicitly says that the structure of the near surface flow is independent of viscosity, v, the rotational frequency of the Earth, f, the depth of the turbulent boundary layer, h, and so on. These arguments are asymptotic rather than absolute statements. They effectively state that the Reynolds number (in this case the inverse of the nondimensional viscosity), $Re = u_*z/v$, is so large that the flow ceases to depend on it. Likewise for the Rossby number, $Ro \equiv u_*/fz$. In these cases, we speak of the flow obeying Reynolds or Rossby number similarity.

The lack of an outer scale measuring the depth of the boundary layer, or the atmosphere as a whole, suggests that, to the extent our idealization is valid, it depends on z being much less than h. Similarly, the application of this formalism assumes that z is much greater than the height of the surface roughness elements. Insofar as all of these statements are true, then α should be universal; that is, once it is empirically determined, it can be universally applied. Given α, the problem of relating the small-scale flux of momentum (which is responsible for accelerating the mean flow) to a function of the mean flow itself is reduced to integration:

$$-\overline{w'u'} = u_*^2 = C_d U^2 \text{ where } C_d \equiv \alpha \ln(z / z_o), \tag{3}$$

where z_0 is an effective height (called the roughness height and defined by the character of the surface) where the extrapolated velocity profile vanishes.

Equation (3) forms the basis for the parameterization of surface fluxes in all models of atmospheric circulations. This approach can be generalized to account for heat fluxes that are accompanied by buoyant acceleration of fluid elements. In this case, α must be replaced by a function $\psi(\zeta)$, where ζ is a nondimensional parameter that measures the relative contributions of buoyancy and mechanically induced effects on fluid accelerations. Further extensions to account for a variety of other effects (most notably surface heterogeneity) not included in the formulation above are invariably also based on elaborations of (3) and remain an active area of research (Fairall et al., 2003).

In concentrating on the details of the formalism above, we should not lose track of the basic ideas of the similarity approach, wherein: (1) insight is used to reduce a problem to an essential and idealized formulation; (2) dimensional analysis is used to identify nondimensional dependencies; and (3) empiricism is used to determine the form of the functions of the relevant dimensionless numbers. Unfortunately, for most processes we wish to parameterize this three-step recipe is not easy to follow. More often than not our insights are not sufficiently developed for us to arrive at compelling simplifications. Even when they are compelling, however, the empirical step often involves measuring functions that involve more than one variable and are not readily accessible to measurement.

To address these problems, a boot-strapping approach has been developed, wherein idealized fluid simulations designed to isolate particular processes, or collections of processes, are used to develop our intuition. Slowly, these simulations are refined to their essence, from which pseudo-empirical statements are extracted, and the parameter space is explored.

An example of this approach is an attempt to parameterize the effects of clouds in large-scale models. Most of the processes directly responsible for cloud formation are related to circulations much smaller than the smallest scale represented by large-scale models. However clouds, and cloud regimes, do exhibit large-scale patterns and thus seem to be under control of the large-scale state. This raises the possibility of using a fine-scale model with a given large-scale forcing to learn which large-scale parameters are essential to cloud formation and how cloud fields respond to changes in these parameters (e.g., Xu and Randall, 1996).

Based on this approach, simple statistical rules can be derived, both for use in larger-scale models and for comparison to data. The latter provides a means of evaluating the fidelity of fine-scale models, which often depend on a parameterization of an even finer-scale process (Stevens and Lenschow, 2001; Randall et al., 2003b). In reference to the initial example, this approach would involve solving directly for the flow over a variety of surfaces. Based on the simulation results, essential parameters would then be isolated, leading to a formulation of the problem similar to Eq. 2. At this stage, simulations for a variety of roughness types could be conducted to evaluate the constancy of α as u_* and z vary.

Given the variety of processes in the atmosphere that are not resolved in large-scale models and our propensity for making problems more complicated rather than simpler, this approach takes enormous effort. Although it can be rewarding when used creatively, it can also often be tedious, particularly when nature resists simplification.

To address these problems, another approach has recently been developed. Here fine-scale simulations are embedded in larger-scale simulations, in a sense performing the procedure outlined above "on the fly." For some processes, this approach has great potential—particularly processes that resist simplification and in situations where myriad interactions among unresolved processes occur on a narrow range of scales that are clearly separated from the smallest of the resolved "large-scales."

Although these qualifications appear onerous, they are satisfied by some elements of one of the more vexing parameterization problems in atmospheric sciences—relating the statistics of deep convective clouds to the state of the large-scale circulations. Recent applications of this approach, called super-parameterization, or cloud-resolving-convective parameterization (Grabowski and Smolarkiewicz, 1999), have led to remarkable improvements in the fidelity of some important aspects of simulations of large-scale atmospheric phenomena (Khairoutdinov and Randall, 2003).

Super-parameterization, however, is very new, and strategies for implementing it are just being explored (Randall et al., 2003a). Computationally, it is very intensive, and thus it requires access to great resources. Nevertheless, it has the potential to enrich our phenomenology and make the more traditional strategies outlined above more effective. For this reason, and because of its immediate practical benefits, it is being explored vigorously.

REFERENCES

Barenblatt, G. 1996. Scaling, Self-Similarity, and Intermediate Asymptotics. Cambridge, U.K.: Cambridge University Press.

Fairall, C.W., E.F. Bradley, J.E. Hare, A. Grachev, and J. Edson. 2003. Bulk parameterization of air-sea fluxes: updates and verification for the COARE algorithm. Journal of Climate 16(4): 571–591.

Garratt, J.R. 1992. The Atmospheric Boundary Layer. Cambridge, U.K.: Cambridge University Press.

Grabowski, W.W., and P.K. Smolarkiewicz. 1999. CRCP: a cloud resolving convective parameterization for modeling the tropical convective atmosphere. Physica D: Nonlinear Phenomena 133(1-4): 171–178.

Khairoutdinov, M.F., and D.A. Randall. 2003. A cloud resolving model as a cloud parameterization in the NCAR Community Climate System Model: preliminary results. Geophysical Research Letters 28(18): 3617–3620.

Randall, D.A., M. Khairoutdinov, A. Arakawa, and W. Grabowski. 2003a. Breaking the cloud parameterization deadlock. Bulletin of the American Meteorological Society 84(11): 1547–1564.

Randall, D.A., S. Krueger, C. Bretherton, J. Currey, P. Duynkerke, M. Moncrieff, B. Ryan, D. Starr, M. Miller, W. Rossow, G. Tselioudis, and B. Wielicki. 2003b. Confronting models with data: the GEWEX Cloud Systems Study. Bulletin of the American Meteorological Society 84(4): 455–469.

Stevens, B., and D.H. Lenschow. 2001. Observations, experiments and large-eddy simulation. Bulletin of the American Meteorological Society 82(2): 283–294.

Xu, K.-M., and D.A. Randall. 1996. A semi-empirical cloudiness parameterization for use in climate models. Journal of Atmospheric Science 53(21): 3084–3102.

ENGINEERING AND ENTERTAINMENT

Introduction

CHRIS KYRIAKAKIS
University of Southern California
Los Angeles, California

Entertainment technology has evolved over the last 100 years from hand-cranked film cameras with no sound to all-digital picture capture and multichannel surround sound. Most of the engineering innovations associated with entertainment have been fueled by the film industry's need to surpass itself every few years. In recent years, entertainment has changed dramatically—from a large group activity available only in a movie theater to an activity that can be enjoyed by a family in a home theater or even an individual with advanced portable entertainment equipment.

Innovations have not only ensured the high quality of the experience but also provided new directions for the creators of entertainment content and the people who enjoy it. The papers in this section focus on three areas of innovation: picture, sound, and actors. At first glance, these three seem to be very traditional. But each them holds a key to advancing entertainment to the next level.

Anyone who has seen a summer blockbuster is aware of the dramatic improvement in computer-generated realism. Visual-effects supervisors now report that that they can bring even the most challenging visions of film directors to the screen. The only questions are time and cost. The technology behind the more realistic computer graphics (CG) techniques is simulations of light traveling in a scene and reflecting off of and through surfaces. These techniques—some developed recently and some originating in the in the 1980s—are being applied to visual-effects processes by CG artists who have found ways to channel the power of these new tools. In "Capturing and Simulating Physically Accurate Illumination in Computer Graphics," Paul Debevec describes how new techniques are bringing unimaginable realism to the screen, creating visual elements that are becoming almost indistinguishable from reality.

Bill Gardner's presentation, "Spatial Audio Reproduction: Toward Individualized Binaural Sound," focuses on spatial perception, a critical aspect of sound reproduction. While the audio industry remains focused on advances that can improve audio quality incrementally, Gardner's work is approaching a new frontier in sound. We perceive the direction, distance, and size of sound sources with our ears. But the accurate reproduction of the spatial properties of sound remains a challenge. In this presentation, the technologies for spatial sound reproduction are reviewed and future directions, with a focus on the promise of individualized binaural technology, are explored.

The third presentation in this section, "Designing Socially Intelligent Robots," by Cynthia Breazeal, addresses advances in entertainment technology, specifically as it applies to robots. Breazeal interprets "entertainment" in a broad sense that encompasses personal-service robots, for which there is a quickly emerging market. Breazeal raises questions about the design of robots that can successfully interact in the daily lives of ordinary people. Beyond performing useful tasks, personal robots must be natural and intuitive for the average consumer to interact, communicate, work with, and teach new skills. To address these challenges, new areas of inquiry, such as human-robot interaction (HRI) and social robotics, are emerging. Social and emotional intelligence will be fundamental to the design of personal-service robots. After all, personal robots should not only be useful to their human users, but people should genuinely enjoy having their robots around.

Capturing and Simulating Physically Accurate Illumination in Computer Graphics

PAUL DEBEVEC
Institute for Creative Technologies
University of Southern California
Marina del Rey, California

Anyone who has seen a recent summer blockbuster has seen the results of dramatic improvements in the realism of computer-generated graphics. Visual-effects supervisors now report that bringing even the most challenging visions of film directors to the screen is no longer in question; with today's techniques, the only questions are time and cost. Based both on recently developed techniques and techniques that originated in the 1980s, computer-graphics artists can now simulate how light travels within a scene and how light reflects off of and through surfaces.

RADIOSITY AND GLOBAL ILLUMINATION

One of the most important aspects of computer graphics is simulating the illumination in a scene. Computer-generated images are two-dimensional arrays of computed pixel values, with each pixel coordinate having three numbers indicating the amount of red, green, and blue light coming from the corresponding direction in the scene. Figuring out what these numbers should be for a scene is not trivial, because each pixel's color is based both on how the object at that pixel reflects light and the light that illuminates it. Furthermore, the illumination comes not only directly from the light sources in the scene, but also indirectly from all of the surrounding surfaces in the form of "bounce" light. The complexities of the behavior of light—one reason the world around us appears rich and interesting—make generating "photoreal" images both conceptually and computationally complex.

As a simple example, suppose we stand at the back of a square white room in which the left wall is painted red, the right wall is painted blue, and the light comes from the ceiling (Figure 1). If we take a digital picture of this room and examine its pixel values, we will indeed find that the red wall is red and the blue wall is blue. But when we look closely at the white wall in front of us, we will notice that it isn't perfectly white. Toward the right it becomes bluish, and toward the left it becomes pink. The reason for this is indirect illumination: toward the right, blue light from the blue wall adds to the illumination on the back wall, and toward the left, red light does the same.

Indirect illumination is responsible for more than the somewhat subtle effect of white surfaces picking up the colors of nearby objects—it is often responsible for most, sometimes all, of the illumination on an object or in a scene. If I sit in a white room illuminated by a small skylight in the morning, the indirect light from the patch of sunlight on the wall lights the rest of the room, not the direct light from the sun itself. If light did not bounce between surfaces, the room would be nearly dark!

In early computer graphics, interreflections of light between surfaces in a scene were poorly modeled. Light falling on each surface was computed solely as a function of the light coming directly from light sources, with perhaps a

FIGURE 1 A simulation of indirect illumination in a scene, known as the "Cornell box." The white wall at the back picks up color from the red and blue walls at the sides. Source: Goral et al., 1984. Reprinted with permission. (The Cornell box may be seen in color at *http://www.graphics.cornell. edu/online/box/ history.html.*)

roughly determined amount of "ambient" light added irrespective of the actual colors of light in the scene. The groundbreaking publication showing that indirect illumination could be modeled and computed accurately was presented at SIGGRAPH 84, when Goral et al. of Cornell University described how they had simulated the appearance of the red, white, and blue room example using a technique known as radiosity.

Inspired by physics techniques for simulating heat transfer, the Cornell researchers first divided each wall of the box into a 7 × 7 grid of patches; for each patch, they determined the degree of its visibility to every other patch, noting that patches reflect less light if they are farther apart or facing away from each other. The final light color of each patch could then be written as its inherent surface color times the sum of the light coming from every other patch in the scene. Despite the fact that the illumination arriving at each patch depends on the illumination arriving (and thus leaving) every other patch, the radiosity equation could be solved in a straightforward way as a linear system of equations.

The result that Goral et al. obtained (Figure 1) correctly modeled that the white wall would subtly pick up the red and blue of the neighboring surfaces. Soon after this experiment, when Cornell researchers constructed such a box with real wood and paint, they found that photographs of the box matched their simulations so closely that people could not tell the difference under controlled conditions. The first "photoreal" image had been rendered!

One limitation of this work was that the time required to solve the linear system increased with the cube of the number of patches in the scene, making the technique difficult to use for complex models (especially in 1984). Another limitation was that all of the surfaces in the radiosity model were assumed to be painted in matte colors, with no shine or gloss.

A subsequent watershed work in the field was presented at SIGGRAPH 86, by Jim Kajiya from Caltech, who published "The Rendering Equation," which generalized the ideas of light transport to any kind of geometry and any sort of surface reflectance. The titular equation of Kajiya's paper stated in general terms that the light leaving a surface in each direction is a function of the light arriving from all directions upon the surface, convolved by a function that describes how the surface reflects light. The latter function, called the bidirectional reflectance distribution function (BRDF), is constant for diffuse surfaces but varies according to the incoming and outgoing directions of light for surfaces with shine and gloss.

Kajiya described a process for rendering images according to this equation using a randomized numerical technique known as path tracing. Like the earlier fundamental technique of ray tracing (Whitted, 1980), path tracing generates images by tracing rays from a camera to surfaces in the scene, then tracing rays out from these surfaces to determine the incidental illumination on the surfaces. In path tracing, rays are traced not only in the direction of light sources, but also randomly in all directions to account for indirect light from the rest of the scene.

FIGURE 2 A synthetic scene rendered using path tracing. Source: Kajiya, 1986. Reprinted with permission.

The demonstration Kajiya produced for his paper is shown in Figure 2. This simple scene shows realistic light interactions among both diffuse and glossy surfaces, as well as other complex effects, such as light refracting through translucent objects. Although still computationally intensive, Kajiya's randomized process for estimating solutions to the rendering equation made the problem tractable both conceptually and computationally.

BRINGING REALITY INTO THE COMPUTER

Using the breakthroughs in rendering techniques developed in the mid-1980s, it was no simple endeavor to produce synthetic images with the full richness and realism of images in the real world. Photographs appear "real" because shapes in the real world are typically distinctive and detailed, and surfaces in the real world reflect light in interesting ways, with different characteristics that vary across surfaces. And also, very importantly, light in the real world is interesting because typically there are different colors and intensities of light coming from every direction, which dramatically and subtly shape the appearance of the forms in a scene. Computer-generated scenes, when constructed from simple shapes, textured with ideal plastic and metallic reflectance properties, and illuminated by simple point and area light sources, lack "realism" no matter how accurate or computationally intensive the lighting simulation. As a result, creating photoreal images was still a matter of skilled artistry rather than advanced technology. Digital artists had to adjust the appearance of scenes manually.

Realistic geometry in computer-generated scenes was considerably advanced in the mid-1980s when 3-D digitizing techniques became available for scanning the shapes of real-world objects into computer-graphics programs. The Cyberware 3-D scanner, an important part of this evolution, transforms objects and human faces into 3-D polygon models in a matter of seconds by moving a stripe of laser light across them. An early use of this scanner in a motion picture was in *Star Trek IV* for an abstract time-travel sequence showing a collage of 3-D models of the main characters' heads. 3-D digitization techniques were also used to capture artists' models of extinct creatures to build the impressive digital dinosaurs for *Jurassic Park*.

DIGITIZING AND RENDERING WITH REAL-WORLD ILLUMINATION

Realism in computer graphics advanced again with techniques that can capture illumination from the real world and use it to create lighting in computer-generated scenes. If we consider a particular place in a scene, the light at that place can be described as the set of all colors and intensities of light coming toward it from every direction. As it turns out, there is a relatively straightforward way to capture this function for a real-world location by taking an image of a mirrored sphere, which reflects light coming from the whole environment toward the camera. Other techniques for capturing omnidirectional images include fisheye lenses, tiled panoramas, and scanning panoramic cameras.

The first and simplest form of lighting from images taken from a mirrored sphere is known as environment mapping. In this technique, the image is directly warped and applied to the surface of the synthetic object. The technique using images of a real scene was used independently by Gene Miller and Mike Chou (Miller and Hoffman, 1984) and Williams (1983). Soon after, the technique was used to simulate reflections on the silvery, computer-generated spaceship in the 1986 film *Flight of the Navigator* and, most famously, on the metallic T1000 "terminator" character in the 1991 film *Terminator 2*. In all of these examples, the technique not only produced realistic reflections on the computer-graphics object, but also made the object appear to have truly been in the background environment. This was an important advance for realism in visual effects. Computer-graphics objects now appeared to be illuminated by the light of the environment they were in (Figure 3).

Environment mapping produced convincing results for shiny objects, but innovations were necessary to extend the technique to more common computer-graphics models, such as creatures, digital humans, and cityscapes. One limitation of environment mapping is that it cannot reproduce the effects of object surfaces shadowing themselves or of light reflecting between surfaces. The reason for this limitation is that the lighting environment is applied directly to the object surface according to its surface orientation, regardless of the degree of visibility of each surface in the environment. For surface points on the convex hull of an object, correct answers can be obtained. However, for more typical

a.

b.

FIGURE 3 Environment-mapped renderings from the early 1980s. **a.** An environment-mapped shiny dog. Source: Miller and Hoffman, 1984. Reprinted with permission. **b.** An environment-mapped shiny robot. Source: Williams, 1993. Reprinted with permission.

points on an object, appearance depends both on which directions of the environment they are visible to and light received from other points on the object.

A second limitation of the traditional environment mapping process is that a single digital or digitized photograph of an environment rarely captures the full range of light visible in a scene. In a typical scene, directly visible light sources are usually hundreds or thousands of times brighter than indirect illumination from the rest of the scene, and both types of illumination must be captured to represent the lighting accurately. This wide dynamic range typically exceeds the dynamic range of both digital and film cameras, which are designed to capture a range of brightness values of just a few hundred to one. As a result, light sources typically become "clipped" at the saturation point of the image sensor, leaving no record of their true color or intensity. This is not a major problem for shiny metal surfaces, because shiny reflections would become clipped anyway in the final rendered images. However, when lighting more typical surfaces—surfaces that blur the incidental light before reflecting it back toward the camera—the effect of incorrectly capturing the intensity of direct light sources in a scene can be significant.

We developed a technique to capture the full dynamic range of light in a scene, up to and including direct light sources (Debevec and Malik, 1997). Photographs are taken using a series of varying exposure settings on the camera; brightly exposed images record indirect light from the surfaces in the scene, and dimly exposed images record the direct illumination from the light sources without clipping. Using techniques to derive the response curve of the imaging system (i.e., how recorded pixel values correspond to levels of scene brightness), we assemble this series of limited-dynamic-range images into a single high-dynamic image representing the full range of illumination for every point in the scene. Using IEEE floating-point numbers for the pixel values of these high-dynamic-range images (called HDR images or HDRIs), ranges exceeding even one to a million can be captured and stored.

The following year we presented an approach to illuminating synthetic objects with measurements of real-world illumination known as image-based lighting (IBL), which addresses the remaining limitations of environment mapping (Debevec, 1998). The first step in IBL is to map the image onto the inside of a surface, such as an infinite sphere, surrounding the object, rather than mapping the image directly onto the surface of the object. We then use a global illumination system (such as the path-tracing approach described by Kajiya [1986]) to simulate this image of incidental illumination actually lighting the surface of the object. In this way, the global illumination algorithm traces rays from each object point out into the scene to determine what is lighting it. Some of the rays have a free path away from the object and thus strike the environmental lighting surface. In this way, the illumination from each visible part of the environment can be accounted for. Other rays strike other parts of the object, blocking the light it would have received from the environment in that direction. If the system computes additional ray bounces, the color of the object at the occluding surface point is computed in a similar way; otherwise, the algorithm approximates the light arriving from this direction as zero. The algorithm sums up all of the light arriving directly and indirectly from the environment at each surface point and uses this sum as the point's illumination. The elegance of this approach is that it produces all of the effects of the real object's appearance illuminated by the light of the environment, including self-shadowing, and it can be applied to any material, from metal to plastic to glass.

We first demonstrated HDRI and IBL in a short computer animation called *Rendering with Natural Light*, shown at the SIGGRAPH 98 computer-animation festival (Figure 4, top). The film featured a still life of diffuse, shiny, and translucent spheres on a pedestal illuminated by an omnidirectional HDRI of the light in the eucalyptus grove at UC Berkeley. We later used our lighting capture techniques to record the light in St. Peter's Basilica, which allowed us to add virtual tumbling monoliths and gleaming spheres to the basilica's interior for our SIGGRAPH 99 film *Fiat Lux* (Figure 4, bottom). In *Fiat Lux*, we used the same lighting techniques to compute how much light the new objects would obstruct from hitting the ground; thus, the synthetic objects cast shadows in the same way they would have if they had actually been there.

The techniques of HDRI and IBL, and the techniques and systems derived from them, are now widely used in the visual-effects industry and have provided visual-effects artists with new lighting and compositing tools that give digital

FIGURE 4 Still frames from the animations *Rendering with Natural Light* (top) and *Fiat Lux* (bottom) showing computer-generated objects illuminated by and integrated into real-world lighting environments.

FIGURE 5 A virtual image of the Parthenon synthetically illuminated with a lighting environment captured at the USC Institute for Creative Technologies.

actors, airplanes, cars, and creatures the appearance of actually being present during filming, rather than added later via computer graphics. Examples of elements illuminated in this way include the transforming mutants in *X-Men* and *X-Men 2*, virtual cars and stunt actors in *The Matrix Reloaded*, and whole cityscapes in *The Time Machine*. In our latest computer animation, we extended the techniques to capture the full range of light of an outdoor illumination environment—from the pre-dawn sky to a direct view of the sun—to illuminate a virtual 3-D model of the Parthenon on the Athenian Acropolis (Figure 5).

APPLYING IMAGE-BASED LIGHTING TO ACTORS

In my laboratory's most recent work, we have examined the problem of illuminating real objects and people, rather than computer-graphics models, with light captured from real-world environments. To accomplish this we use a series of light stages to measure directly how an object transforms incidental environmental illumination into reflected radiance, a data set we call the reflectance field of an object.

The first version of the light stage consisted of a spotlight attached to a two-bar rotation mechanism that rotated the light in a spherical spiral about a person's face in approximately one minute (Debevec et al., 2000). At the same time, one or more digital video cameras recorded the object's appearance under every form of directional illumination. From this set of data, we could render the object under any form of complex illumination by computing linear combinations of the color channels of the acquired images. The illumination could be chosen to be measurements of illumination in the real world (Debevec, 1998) or the illumination present in a virtual environment, allowing the image of a real person to be photorealistically composited into a scene with the correct illumination.

An advantage of this photometric approach for capturing and rendering objects is that the object need not have well defined surfaces or easy-to-model reflectance properties. The object can have arbitrary self-shadowing, interreflection, translucency, and fine geometric detail. This is helpful for modeling and rendering human faces, which exhibit all of these properties, as do many objects we encounter in our everyday lives.

Recently, our group constructed two additional light stages. Light Stage 2 (Figure 6) uses a rotating semicircular arm of strobe lights to illuminate the face from a large number of directions in about eight seconds, much more quickly than Light Stage 1. For this short a period of time, an actor can hold a steady facial expression for the entire capture session. By blending the geometry and reflectance of faces with different facial expressions, we have been able to create novel animated performances that can be realistically rendered from new points of view and under arbitrary illumination (Hawkins et al., 2004). Mark Sagar and his colleagues at Sony Pictures Imageworks used related techniques to create the digital stunt actors of Tobey Maguire and Alfred Molina from light stage data sets for the film *Spider-Man 2*.

For Light Stage 3, we built a complete sphere of 156 light sources that can illuminate an actor from all directions simultaneously (Debevec et al., 2002). Each light consists of a collection of red, green, and blue LEDs interfaced to a computer so that any light can be set to any color and intensity. The light stage

FIGURE 6 Light Stage 2 is designed to illuminate an object or a person from all possible directions in a short period of time, allowing a digital video camera to capture directly the subject's reflectance field. Source: Hawkins et al., in press. Reprinted with permission.

can be used to reproduce the illumination from a captured lighting environment by using the light stage as a 156-pixel display device for the spherical image of incidental illumination. A person standing inside the sphere then becomes illuminated by a close approximation of the light that was originally captured. When composited over a background image of the environment, it appears nearly as if the person were there. This technique may improve on how green screens and virtual sets are used today. Actors in a studio can be filmed lit as if they were somewhere else, giving visual-effects artists much more control over the realism of the lighting process.

In our latest tests, we use a high-frame camera to capture how an actor appears under several rapidly cycling basis lighting conditions throughout the course of a performance (Debevec et al., 2004). In this way, we can simulate the actor's appearance under a wide variety of different illumination conditions after filming, providing directors and cinematographers with never-before-available control of the actor's lighting during postproduction.

A REMAINING FRONTIER: DIGITIZING REFLECTANCE PROPERTIES

Significant challenges remain in the capture and simulation of physically accurate illumination in computer graphics. Although techniques for capturing object geometry and lighting are maturing, techniques for capturing object reflectance properties—the way the surfaces of a real-world object respond to light—are still weak. In a recent project, our laboratory presented a relatively simple technique for digitizing surfaces with varying color and shininess components (Gardner et al., 2003). We found that by moving a neon tube light source across a relatively flat object and recording the light's reflections using a video camera we could independently estimate the diffuse color and the specular properties of every point on the object. For example, we digitized a 15th-century illuminated manuscript with colored inks and embossed gold lettering (Figure 7a). Using the derived maps for diffuse and specular reflection, we were able to render computer-graphics versions of the manuscript under any sort of lighting environment with realistic glints and reflections from different object surfaces.

A central complexity in digitizing reflectance properties for more general objects is that the way each point on an object's surface responds to light is a complex function of the direction of incidental light and the viewing direction—the surface's four-dimensional BRDF. In fact, the behavior of many materials and objects is even more complicated than this, in that incidental light on other parts of the object may scatter within the object material, an effect known as subsurface scattering (Hanrahan and Krueger, 1993). Because this effect is a significant component of the appearance of human skin, it has been the subject of interest in the visual-effects industry. New techniques for simulating subsurface scattering effects on computer-generated models (Jensen et al., 2001) have led to more realistic renderings of computer-generated actors (Figure 7b) and creatures, such as the Gollum character in *Lord of the Rings*.

Obtaining models of how real people and objects scatter light in their full generality is a subject of ongoing research. In a recent study, Goesele et al. (2004) used a computer-controlled laser to shine a narrow beam onto every point of a translucent alabaster sculpture (Figure 7c) and recorded images of the resulting light scattering using a specially chosen high-dynamic-range camera. By making the simplifying assumption that any point on the object would respond equally to any incidental and radiant light direction, the dimensionality of the problem was reduced from eight to four dimensions yielding a full characterization of the object's interaction with light under these assumptions. As research in this area continues, we hope to develop the capability of digitizing anything—no matter what it is made of or how it reflects light—so it can become an easily manipulated, photoreal computer model. For this, we will need new acquisition and analysis techniques and continued increases in computing power and memory capacity.

ACKNOWLEDGMENTS

The author wishes to thank Andrew Gardner, Chris Tchou, Tim Hawkins, Andreas Wenger, Andrew Jones, Maya Martinez, David Wertheimer, and Lora Chen for their support during the preparation of this article. Portions of the work described in this article have been sponsored by a National Science Foundation Graduate Research Fellowship, a MURI Initiative on three-dimensional direct visualization from ONR and BMDO (grant FDN00014-96-1-1200), Interval Research Corporation, the University of Southern California, and U.S. Army contract number DAAD19-99-D-0046. Any opinions, findings, and conclusions or recommendations expressed in this paper do not necessarily reflect the views of the sponsors. More information on many of these projects is available at <*http://www.debevec.org/*>.

REFERENCES

Debevec, P. 1998. Rendering synthetic objects into real scenes: bridging traditional and image-based graphics with global illumination and high dynamic range photography. Pp. 189–198 of the Proceedings of the Association for Computing Machinery Special Interest Group on Computer Graphics and Interactive Techniques 98 International Conference. New York: ACM Press/ Addison-Wesley.

Debevec, P., A. Gardner, C. Tchou, and T. Hawkins. 2004. Postproduction re-illumination of live action using time-multiplexed lighting. Technical Report No. ICT TR 05.2004. Los Angeles: Institute for Creative Technologies, University of Southern California.

a.

b.

c.

FIGURE 7 **a.** A digital model of a digitized illuminated manuscript lit by a captured lighting environment. Source: Gardner et al., 2003. Reprinted with permission. **b.** A synthetic model of a face using a simulation of subsurface scattering. Source: Jensen et al., 2001. Reprinted with permission. **c.** A digital model of a translucent alabaster sculpture. Source: Goesele et al., 2004. Reprinted with permission.

Debevec, P., T. Hawkins, C. Tchou, H.P. Duiker, W. Sarokin, and M. Sagar. 2000. Acquiring the reflectance field of a human face. Pp. 145–156 of the Proceedings of the Association for Computing Machinery Special Interest Group on Computer Graphics and Interactive Techniques 97 International Conference. New York: ACM Press/Addison-Wesley.

Debevec, P.E., and J. Malik. 1997. Recovering high dynamic range radiance maps from photographs. Pp. 369–378 of the Proceedings of the Association for Computing Machinery Special Interest Group on Computer Graphics and Interactive Techniques 97 International Conference. New York: ACM Press/Addison-Wesley.

Debevec, P., A. Wenger, C. Tchou, A. Gardner, J. Waese, and T. Hawkins. 2002. A lighting reproduction approach to live-action compositing. Association for Computing Machinery Transactions on Graphics 21(3): 547–556.

Gardner, A., C. Tchou, T. Hawkins, and P. Debevec. Linear light source reflectometry. 2003. Association for Computing Machinery Transactions on Graphics 22(3): 749–758.

Goesele, M., H.P.A. Lensch, J. Lang, C. Fuchs, and H-P. Seidel. 2004. DISCO—Acquisition of Translucent Objects. Pp. 835–844 of the Proceedings of the Association for Computing Machinery Special Interest Group on Computer Graphics and Interactive Techniques 31st International Conference, August 8–14, 2004.

Goral, C.M., K.E. Torrance, D.P. Greenberg, and B. Battaile. 1984. Modelling the interaction of light between diffuse surfaces. Association for Computer Machinery Computer Graphics (3): 213–222.

Hanrahan, P., and W. Krueger. 1993. Reflection from layered surfaces due to subsurface scattering. Pp. 165–174 of the Proceedings of the Association for Computing Machinery Special Interest Group on Computer Graphics and Interactive Techniques 1993 International Conference. New York: ACM Press/Addison-Wesley.

Hawkins, T., A. Wenger, C. Tchou, A. Gardner, F. Goransson, and P. Debevec. 2004. Animatable facial reflectance fields. Proceedings of the 15th Eurographics Symposium on Rendering in Norrkoping, Sweden, June 21–23, 2004.

Jensen, H.W., S.R. Marschner, M. Levoy, and P. Hanrahan. 2001. A practical model for subsurface light transport. Pp. 511–518 of the Proceedings of the Association for Computing Machinery Special Interest Group on Computer Graphics and Interactive Techniques 2001 International Conference. New York: ACM Press/Addison-Wesley.

Kajiya, J.T. 1986. The rendering equation. Association for Computing Machinery Special Interest Group on Computer Graphics and Interactive Techniques Computer Graphics 20(4): 143–150.

Miller, G.S., and C.R. Hoffman. 1984. Illumination and Reflection Maps: Simulated Objects in Simulated and Real Environments. Course Notes for Advanced Computer Graphics Animation. New York: Association for Computing Machinery Special Interest Group on Computer Graphics and Interactive Techniques Computer Graphics.

Whitted, T. 1980. An improved illumination model for shaded display. Communications of the Association for Computing Machinery 23(6): 343–349.

Williams, L. 1983. Pyramidal parametrics. 1983. Association for Computing Machinery Special Interest Group on Computer Graphics and Interactive Techniques Computer Graphics 17(3): 1–11.

Spatial Audio Reproduction:
Toward Individualized Binaural Sound

WILLIAM G. GARDNER
Wave Arts, Inc.
Arlington, Massachusetts

The compact disc format, which records audio with 16-bit resolution at a sampling rate of 44.1 kHz, was engineered to reproduce audio with fidelity exceeding the limits of human perception. And it works. However, sound is inherently a spatial perception. We perceive the direction, distance, and size of sound sources, and reproducing the spatial properties of sound accurately remains a challenge. In this paper, I review the technologies for spatial sound reproduction and discuss future directions, focusing on the promise of individualized, binaural technology.

HEARING

People hear with two ears, and the two audio signals received at the eardrums completely define the auditory experience. An amazing feature of the auditory system is that with only two ears sounds can be perceived from all directions, and the listener can even sense the distance and size of sound sources. The perceptual cues for sound localization include the amplitude of the sound at each ear, the arrival time at each ear, and the spectrum of the sound, that is, the relative amplitude of the sound at different frequencies.

The spectrum of a sound is modified by interactions between sound waves and the torso, head, and external ear (pinna). Furthermore, spectral modification depends on the location of the source in a complex way. The auditory system uses spectral modifications as cues to the location of sound, but because the complex shape of the pinna varies significantly among individuals, the cues for sound localization are idiosyncratic. Each individual's auditory system is adapted to the idiosyncratic spectral cues produced by his or her head features.

BINAURAL AUDIO

Binaural audio refers specifically to the recording and reproduction of sound at the ears. Binaural recordings can be made by placing miniature microphones in the ear canals of a human subject. Exact reproduction of the recording is possible through properly equalized headphones. If the recording and playback are for the same subject and there are no head movements, the results are stunningly realistic.

Many virtual-reality audio applications attempt to position a sound arbitrarily around a listener wearing headphones. They rely on a stored database of head-related transfer functions (HRTFs), that is, mathematical descriptions of the transformation of sound by the torso, head, and external ear. HRTFs for the left and right ears of a subject specify how sound from a particular direction is transformed en route to the ear drums. A complete description of a subject's head response requires hundreds of HRTF measurements from all directions surrounding the subject. Any sound source can be virtually located by filtering the sound with the HRTFs corresponding to the desired location and presenting the resulting binaural signal to the subject using properly equalized headphones. When this procedure is individualized by using the subject's own HRTFs, the localization performance is equivalent to free-field listening (Wightman and Kistler, 1989a,b).

Figure 1 shows the magnitude spectra for right ear HRTFs measured for three different human subjects with a sound source located on the horizontal plane at 60 degrees right azimuth. Note that the spectra are similar up to 6 kHz; the significant differences in HRTFs at higher frequencies are attributable to variations in pinna shape. Figure 2 shows the magnitude spectra of HRTFs measured from a dummy head microphone for all locations on the horizontal plane. Note how the spectral features change as a function of source direction.

Most research has been focused on localization; subjects presented with an acoustic stimulus are asked to report the apparent direction. Their localization is then compared to free-field listening to assess the quality of reproduction. But this method does not account for many attributes of sound perception, including distance, timbre, and size. In an experimental paradigm developed by Hartmann and Wittenburg (1996), the virtual stimulus is reproduced using open-air headphones that allow free-field listening. Thus, real and virtual stimuli can be compared directly. In these experiments, subjects are presented with a stimulus and asked to decide if it is real or virtual. If a virtual stimulus cannot be distinguished from a real stimulus, then the reproduction error is within the limits of perception. When this experimental paradigm was used to study the externalization of

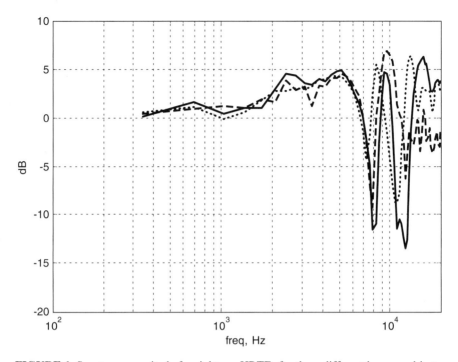

FIGURE 1 Spectrum magnitude for right-ear HRTFs for three different human subjects with a sound source at 60 degrees right azimuth on the horizontal plane. Due to variations in ear shape, the HRTFs differ significantly above 6 kHz.

virtual sound, the results demonstrated that individualized spectral cues are necessary for proper externalization.

The major limitation of binaural techniques is that all listeners are different. Binaural signals recorded for subject A may not sound correct to subject B. Nevertheless, by necessity, binaural systems are seldom individualized. Instead, a reference head, often a model that represents a typical listener or HRTFs known to perform adequately for a range of different listeners, is used to encode binaural signals for all listeners. This is called a "non-individualized" system (Wenzel et al., 1993).

The use of non-individualized HRTFs is limited by a lack of externalization (the sounds are localized in the head or very close to the head), incorrect perception of elevation angle, and front/back reversals. Externalization can be improved somewhat by adding dynamic head tracking and reverberation. Nevertheless, the lack of realistic externalization is often cited as a problem with these systems.

The great challenge in binaural technology is to devise a practical method by which binaural signals can be individualized to a specific listener. There are several possible approaches to meeting this challenge: acoustic measurement,

(a) ipsilateral

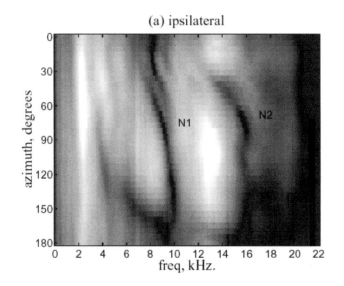

(b) contralateral

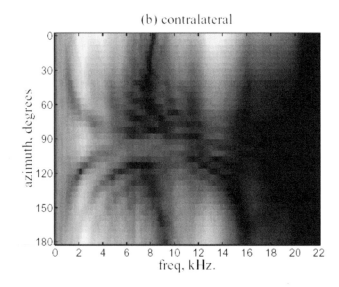

FIGURE 2 Magnitude spectra of KEMAR dummy-head HRTFs as a function of azimuth for a horizontal source. Source: Gardner, 1998. Reprinted with permission. **a.** Ipsilateral (same side) ear. **b.** Contralateral (opposite side) ear. White indicates +10 dB; black indicates –30 dB. Notch features are labeled in (a) according to Lopez-Poveda and Meddis (1996). The figure shows the complex, yet systematic, variation in spectrum as a function of source direction.

statistical models, calibration procedure, simplified geometrical models, and accurate head models solved using computational acoustics.

With the proper equipment, measuring the HRTFs of a listener is a straightforward procedure, although not practical for commercial applications. Microphones are placed in the ears of the listener, either probe microphones placed somewhere in the ear canal or microphones that block the entrances to the ear canals. Measurement signals are produced from speakers surrounding the listener to measure the impulse response of each source direction to each ear. Because tens or hundreds of directions may be measured, the listener is positioned either in a rotating chair or in a fixed position surrounded by hundreds of speakers. The measurements are often made in an anechoic (echo-free) chamber.

Various statistical methods have been used to analyze databases of HRTF measurements in an effort to tease out some underlying structure in the data. One important study applied principal component analysis (PCA) to a database of HRTFs from 10 listeners at 256 directions (Kistler and Wightman, 1992). Using the log magnitude spectra of the HRTFs as input, the analysis indicated that 90 percent of the variance in the data could be accounted for using only five principal components. The study tested the localization performance using individualized HRTFs approximated by weighted sums of the five principal components. When the listener's own HRTFs were used, the results were nearly identical. The study gathered only directional judgments, and externalization was not considered. The study showed that a five-parameter model is sufficient for synthesizing individualized HRTF spectra, at least in terms of directional localization and for a single direction. Unfortunately, the five parameters must be calculated for each source direction, which means individualized measurements are still necessary.

One can imagine a simple calibration procedure that would involve the listener adjusting knobs to match a parameterized HRTF model with the listener's characteristics. The listener could be given a test stimulus and asked to adjust a knob until some attribute of his perception was maximized. After adjusting several knobs in this manner, the parameter values of the internal model would be optimized for the listener, and the model would be able to generate individualized HRTFs for that individual. Some progress has been made in this area. For example, it has been demonstrated that calibrating HRTFs according to overall head size improves localization performance (Middlebrooks et al., 2000). However, to date, detailed methods of modeling and calibrating the data have not been found.

Many researchers have developed geometrical models for the torso, head, and ears. The head and torso can be modeled using ellipsoids (Algazi and Duda, 2002), and the pinna can be modeled as a set of simple geometrical objects (Lopez-Poveda and Meddis, 1996). For simple geometries, the acoustic-wave equation can be solved to determine head response. For more complicated geometries, head response can be approximated using a multipath model, wherein each reflecting or diffracting object contributes an echo to the response (Brown and Duda, 1997). In theory, head models should be easy to fit to any particular listener by making anthropometric measurements of the listener and plugging these into the model. However, studies have shown that although simplified geometrical models are accurate at low frequencies, they become increasingly inaccurate at higher frequencies. Because of the importance of high-frequency localization cues for proper externalization, elevation perception, and front/back resolution of sound, simplified geometrical models are not suitable for creating individualized HRTFs.

A more promising approach has been to use a three-dimensional laser scan to produce an accurate geometrical representation of a head as a basis for computational acoustic simulation using finite-element modeling (FEM) or boundary-element modeling (BEM) (Kahana et al., 1998, 1999). With this method, HRTFs can be determined computationally with the same accuracy as acoustical measurements, even at high frequencies. Using a 15,000-element model of the head and ear (Figure 3), Kahana demonstrated computation of HRTFs that match acoustical measurements very precisely up to 15 kHz.

There are, however, a number of practical difficulties with this method.

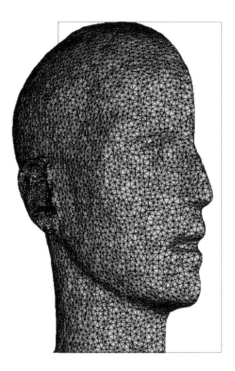

FIGURE 3 Mesh model of one-half of a KEMAR dummy head using 15,000 elements. The model is sufficiently detailed to produce HRTFs that match acoustical measurements, even at high frequencies. Source: Kahana, 1999. Reprinted with permission.

First, scanning the head is complicated by the presence of hair, obscured areas behind the ears, and the obscured internal features of the ear. Second, replicating the interior features of the ear requires making molds and then scanning them separately. Third, after the various scans are spliced together, the number of elements in the model must be pruned to computationally tractable quantities without compromising spatial resolution. Finally, solution of the acoustical equations requires significant computation. For all of these reasons, this approach currently requires more effort and expense than acoustical measurement of HRTFs.

The technique does suggest an alternative approach to determining individualized HRTFs. A deformable head model could be fashioned from finite elements and parameterized with a set of anthropometric measurements. After making head measurements of a particular subject and plugging these into the model, the model head would "morph" into a close approximation of the subject's head. At that point, the computational acoustics procedure could be used to determine individualized HRTFs for the subject. Ideally, the subject's measurements could be determined from images using computer vision techniques. The goal would be a system that could automatically determine individualized HRTFs based on a few digital images of the subject's head and ears. The challenges will be to develop a head model that can morph to fit any head, to obtain a sufficiently accurate ear shape, and to develop ways to estimate the parameters from images of the subject.

CROSS-TALK–CANCELLED AUDIO

Binaural audio can be delivered to a listener over conventional stereo loudspeakers, but each loudspeaker (unlike headphones) creates significant "crosstalk" to the opposite ear. The cross-talk can be cancelled by preprocessing the speaker signals (called cross-talk cancellers) with the inverse of the 2×2 matrix of transfer functions from the speakers to the ears. Cross-talk cancellers use a model of the head to anticipate what cross-talk will occur, then add an out-of-phase cancellation signal to the opposite channel. Thus, the cross-talk is acoustically cancelled at the listener's ears. If the head responses of the listener are known, and if the listener's head remains fixed, an individualized cross-talk cancellation system can be designed that works extremely well.

Non-individualized systems are effective only up to 6 kHz and then only when the listener's position is known (Gardner, 1998). However, despite poor high-frequency performance, cross-talk–cancelled audio is capable of producing stunning, well externalized, virtual sounds to the sides of the listener using frontally placed loudspeakers. As a result of the listener's pinna cues, the sounds are well externalized. The sounds are shifted to the side as a result of the dominance of low-frequency time-delay cues in lateral localization; the cross-talk cancellation works effectively at low frequencies to provide this cue.

MULTICHANNEL AUDIO

The first audio reproduction systems were monophonic, reproducing a single audio signal through one transducer. Stereophonic audio systems, recording and reproducing two independent channels of audio, sound much more realistic. With two loudspeakers, it is possible to position a sound source at either speaker or to position sounds between the speakers by sending a proportion of the sound to each speaker. Stereo has a great advantage over mono because it reproduces a set of locations between the speakers. Also with stereo, uncorrelated signals can be sent to the two ears, which is necessary to achieve a sense of space.

Multichannel audio systems, such as the current 5.1 surround systems, have continued the trend of adding channels around the listener to improve spatial reproduction. 5.1 systems have left, center, and right frontal speakers, left and right surround speakers positioned to the sides of the listener, and a subwoofer to reproduce low frequencies. Because 5.1 systems were designed for cinema sound, the focus is on accurate frontal reproduction so that movie dialogue is spatially aligned with images of the actors speaking. The surround speakers are used for off-screen sounds or uncorrelated ambient effects. The trend in multichannel audio is to add more speaker channels to improve the accuracy of on-screen sounds and provide additional locations for off-screen sounds. As increasing numbers of speakers are added at the perimeter of the listening space, it becomes possible to reconstruct arbitrary sound fields within the space, a technology called wave-field synthesis.

ULTRASONIC AUDIO

Ultrasonics can be used to produce highly directional audible sound beams. This technology is based on physical properties of air, particularly that air becomes a nonlinear medium at high sound pressures. Hence, it is possible to transmit two high-intensity ultrasonic tones, say at 100 kHz and 101 kHz, and produce an audible 1 kHz tone as a result of the intermodulation between the two ultrasonic tones. However, the demodulated signal will be significantly distorted, so the audio must be preprocessed to reduce the distortion after demodulation (Pompei, 1999). Although this technology is impressive, it cannot reproduce low-frequency sounds effectively, and it has lower fidelity than standard loudspeakers.

SUMMARY

Binaural audio has the potential to reproduce sound that is indistinguishable from sounds in the real world. However, the playback must be individualized to each listener's head response. This is currently possible by making acoustical measurements or by making geometrical scans and applying computational acoustic modeling. A practical means of individualizing head responses has yet to be developed.

REFERENCES

Algazi, V.R., and R.O. Duda. 2002. Approximating the head-related transfer function using simple geometric models of the head and torso. Journal of the Acoustical Society of America 112(5): 2053–2064.

Brown, C.P., and R.O. Duda. 1997. An efficient HRTF model for 3-D sound. Pp. 298–301 in Proceedings of the IEEE Workshop on Applications of Signal Processing to Audio and Acoustics. New York: IEEE.

Gardner, W.G. 1998. 3-D Audio Using Loudspeakers. Boston, Mass.: Kluwer Academic Publishers.

Hartmann, W.M., and A. Wittenberg. 1996. On the externalization of sound images. Journal of the Acoustical Society of America 99(6): 3678–3688.

Kahana, Y., P.A. Nelson, and M. Petyt. 1998. Boundary element simulation of HRTFs and sound fields produced by virtual acoustic imaging. Proceedings of the Audio Engineering Society's 105th Convention. Preprint 4817, unpaginated.

Kahana, Y., P.A. Nelson, M. Petyt, and S. Choi. 1999. Numerical modeling of the transfer functions of a dummy-head and of the external ear. Pp. 330–334 in Proceedings of the Audio Engineering Society's 16th International Conference. New York: Audio Engineering Society.

Kistler, D.J., and F.L. Wightman. 1992. A model of head-related transfer functions based on principal components analysis and minimum-phase reconstruction. Journal of the Acoustical Society of America 91(3): 1637–1647.

Lopez-Poveda, E.A., and R. Meddis. 1996. A physical model of sound diffraction and reflections in the human concha. Journal of the Acoustical Society of America 100(5): 3248–3259.

Middlebrooks, J.C., E.A. Macpherson, and Z.A. Onsan. 2000. Psychophysical customization of directional transfer functions for virtual sound localization. Journal of the Acoustical Society of America 108(6): 3088–3091.

Pompei, F.J. 1999. The use of airborne ultrasonics for generating audible sound beams. Journal of the Audio Engineering Society 47(9): 726–731.

Wenzel, E.M., M. Arruda, D.J. Kistler, and F.L. Wightman. 1993. Localization using nonindividualized head-related transfer functions. Journal of the Acoustical Society of America 94(1): 111–123.

Wightman, F.L., and D.J. Kistler. 1989a. Headphone simulation of free-field listening I: stimulus synthesis. Journal of the Acoustical Society of America 85(2): 858–867.

Wightman, F.L., and D.J. Kistler. 1989b. Headphone simulation of free-field listening II: psychophysical validation. Journal of the Acoustical Society of America 85(2): 868–878.

Designing Socially Intelligent Robots

CYNTHIA BREAZEAL
Media Arts and Sciences
Massachusetts Institute of Technology
Cambridge, Massachusetts

The emerging market for personal-service robots raises questions about the design of robots that can play a role in the daily lives of ordinary people. Beyond performing useful tasks, average consumers want personal robots to be natural, intuitive beings with which they can interact, communicate, work with as partners, and teach new skills, knowledge, and tasks (see Fong et al., 2003, for a review). Human-robot interaction (HRI) and social robotics are emerging areas of inquiry in the field of autonomous robotics. In this paper, I argue that social and emotional intelligence will be fundamental to the design of personal-service robots (Breazeal, 2002). After all, personal robots should not only be useful to their human users, but humans should also genuinely enjoy having their robots around.

HISTORICAL PRECURSORS

The idea of creating lifelike robots has amused and fascinated people for thousands of years. Throughout history, humans have attempted to mimic the appearance, functionality, and longevity, as well as the cognitive and adaptive processes, of biological creatures. The idea of lifelike machines appears in Homer's *Iliad* when Hephaistos, the god of metalsmiths, fashions mechanical helpers—strong, vocal, intelligent maidens of gold. The idea surfaces again in medieval times in the Jewish legend of the Golem, a robot-like servant made of clay brought to life by Rabbi Loew to save the Jews of Prague.

As technology advanced, people actually began to build such machines. The first technological breakthrough occurred in the 15th century with mechanical clocks. One hundred years later, clock makers began to build mechanical animals. There is even some evidence that as early as 1478, the young Leonardo da Vinci conceptualized a humanoid automaton controllable by a very crude but programmable analog computer composed of cogs and pulleys (Rosheim, 2000). Nearly 40 years later, in 1515, Leonardo built his famous self-propelled mechanical lion, commissioned by the Medici, which reportedly walked from its place in the room, opened its breast full of lilies, and presented them as a token of friendship from the Medici to Francis I, King of France. In response to the 18th century craze for animated objects, Jacques de Vaucanson created the famous mechanical duck in 1738 that could flap its wings, eat, and digest grain (which still remains a mystery) (Doyon and Liaigre, 1966). In the 1830s or 1840s, Joseph Faber invented a mechanical talking head, called Euphonia, which reputedly could be made to speak in several European languages (Lindsay, 1997). These are just a few of many examples of historical mechanical automata; a more complete account can be found in Rosheim (1994).

The year 1946 marks the invention of the ENIAC computer, the first large-scale, general-purpose, electronic digital computer (McCartney, 1999). Just a few years later, in 1950, the famous British mathematician, Alan Turing, wrote a provocative paper called, "Computing Machinery and Intelligence," in which he discussed the possibility of building machines that can think and learn. Turing outlines a test (the "imitation game," later known as the Turing Test) to determine if a machine can think (Turing, 1950). That same year, Grey Walter published his work on building two robotic tortoises out of analog circuitry that could navigate towards a light source and interact with one another in simple ways (Walter, 1950). In the science fiction arena, Isaac Asimov published his famous three laws of robotics (Asimov, 1942). A visionary Walt Disney applied robotic technology to entertainment for the earliest physically animated performers, such as the famous Abraham Lincoln audio-animatronic that debuted at the 1964 New York World's Fair.

MODERN APPLICATIONS

Today robotic technology is used in entertainment for many purposes. We are familiar with animatronics in theme parks and the use of sophisticated robotic puppets for special effects in films. With recent advances in low-cost electronics, robots can now interact with people in an entertaining, engaging, or anthropomorphic way. In fact, interacting with people has become an important aspect of a robot's functionality. For instance, a new generation of robotic toys has emerged—many of them inexpensive, but some are more expensive and rather sophisticated, such as Sony's robotic dog, Aibo. Robotic kits for "edutainment," such as Lego's Mindstorms, allows kids and adults alike to create their own robotic inventions.

Location-based entertainment robotics, such as robotic museum tour guides, not only entertain visitors, but also provide them with information (Nourbakhsh et al., 1999). Health-related applications are being explored, such as robotic pet-therapy surrogates intended to provide the same health benefits as their living counterparts. Even robots for scientific purposes are beginning to have more socially interactive qualities. For instance, NASA Johnson Space Center's humanoid robot, Robonaut, is ultimately envisioned to be a completely autonomous astronaut's assistant that can work as a productive and cooperative member of a human-robot team (Bluethman et al., 2003).

What about the science-fiction dream of having your very own Star Wars R2-D2 or C-3PO—an appealing robotic sidekick that helps you in your daily life? We are starting to see precursors of such futuristic visions in university and corporate research labs around the world, such as Honda's humanoid robot, ASIMO. Toyota recently announced the Partner Robot Project, which has a stated goal of developing humanoid robots that function as personal assistants for humans. These robots shall "have human characteristics, such as being agile, warm and kind and also intelligent enough to skillfully operate a variety of devices in the area of personal assistance, care for the elderly, manufacturing and mobility."

Robotic Trends magazine defines personal-service robots as "robots or robotic technology purchased by individual consumers that educate, entertain, assist, or protect in the home." One of the strongest motivations for the development of personal robots is to provide domestic assistance and care for the elderly. The global demographic trend of rapidly aging societies, in which a smaller working-age population is responsible for supporting a larger retired population, has created an urgent need for robots that can be capable assistants for people in their homes and can supplement the workforce.

The International Monetary Fund predicts that Japan, in particular, will experience a dramatic change in the ratio of working-age people to retired-age people—from 4:1 today to 2:1 by 2025. In addition, the convergence of many technological developments in mobile computing, such as advances in microprocessor technology, wireless technology, image processing, speech recognition, motor-sensor technology, and embedded systems development tools, have made the development of personal robots increasingly feasible.

Although the service-robot market is still immature, the few quantitative studies that have been done indicate that the market for personal-service robots is on the verge of dramatic growth. Recent research by the Japan Robotics Association (JRA), United Nations Economic Commission (UNEC), and International Federation of Robotics (IFR) indicates that the service-robot market will experience exceptional growth, both in the near term (from $600 million in 2002 to approximately $6 billion in 2009) and for the next few decades (reaching an estimated $60 billion by 2025) (UNECE and IFR, 2002). Of course, one must always take extrapolations from existing studies about the future of immature markets with a large grain of salt. Nevertheless, if these predictions are correct, personal robots will be a ubiquitous technology.

THE PSYCHOLOGY OF ROBOT DESIGN

The success of personal-service robots depends not only on their utility, but also on their ability to be responsive to and interact with ordinary people in a natural and intuitive way. Furthermore, because they may coexist with people on a daily basis, their long-term appeal will certainly affect our willingness to accept them into our lives. For instance, longitudinal studies on the adoption and impact of assistive technologies for the elderly have shown that functionality and need are only part of the design equation. Social and emotional factors also greatly affect the individual's willingness to adopt the technology. Technologies that are stigmatizing (i.e., that make the user feel feeble or vulnerable or make the user feel that they appear that way to others) are often rejected. Even worse, if stigmatizing technologies are adopted, they can contribute to self-imposed isolation or depression (Forlizzi et al., 2004). Thus designing personal robots that support humans socially and emotionally will be just as important as designing them for their cognitive abilities.

According to *The Design of Everyday Things*, in order for people to interact with another entity, they must have a good conceptual model of how that entity operates, whether it is a device, a robot, or even another person (Norman, 1990). If they have such a model, people can explain and predict what an entity may do, understand the reasons for doing it, and know how to elicit desired behavior. The design of a technological artifact, whether it is a robot, a computer, or a teapot, can help a person form this model by "projecting an image of its operation," either through visual cues or continual feedback. By adhering to natural signals and mappings (e.g., physical metaphors or social norms), the artifact becomes intuitively understandable to people.

Numerous HCI studies suggest that people apply a social model when observing and interacting with autonomous robots (Kiesler and Goetz, 2002). Studies by Reeves and Nass (1996) have shown that people treat even desktop computers as social entities and adhere to social norms in their interactions with them. In fact, studies demonstrate that it takes surprisingly few cues to elicit social behavior—a text interface alone is sufficient. Autonomous robots, of course, are quite different from desktop computers in their projected animacy. Like the behavior of living things, the behavior of autonomous robots is a product of their internal state, as well as physical laws. They perceive the world, make decisions, and perform coordinated actions to carry out tasks. If this self-directed, creature-like behavior can be augmented by an ability to communicate with, cooperate with, and learn from people, people will be encouraged to anthropomorphize them. This is true, even for simple vehicles, such as those described in Braitenberg, 1984.

Social robots are a class of autonomous robots explicitly designed to encourage people to socially interact with and understand them. If social robots have personalities, people may be more likely to have a good mental model for them. According to Norman (2004), personality is a powerful design tool for helping people form a conceptual model that channels beliefs, behavior, and intentions in a cohesive and consistent set of behaviors. From a design perspective, the emotion system of a robot could implement the style and personality of the robot, encoding and conveying its attitudes and behavioral inclinations toward the events it encounters. The robot's personality must be designed so that its behavior is understandable and predictable to people. Therefore, parameters of the personality must fall within recognizable human (or animal) norms; otherwise, the robot may appear to be mentally ill or completely alien. The science of natural behavior, as well as artistic insights from classical animation and character design (Thomas and Johnson, 1981), can be useful guides to design in this respect.

ROBOTS WITH SOCIAL AND EMOTIONAL INTELLIGENCE

As robot designers, we tend to emphasize the cognitive aspect of intelligence when designing robot architectures; we tend to view the social, especially the emotional aspect with skepticism (see Sloman and Croucher, 1980, for an exception). However, numerous scientific studies continue to reveal the reciprocal roles of cognition and emotion in intelligent decision making, planning, learning, attention, communication, social interaction, memory, and more (see Isen, 2000, for a review). Cognition and emotion are conceptually distinct, complementary information-processing systems that evolved in response to social and environmental pressures to ensure the health and optimal functioning of the creature (Damasio, 1994). As Norman et al. (2003) argue, the cognitive system is responsible for interpreting and making sense of the world; the emotional system is responsible for evaluating and judging events to assess their overall value with respect to the creature (e.g., positive or negative, desirable or undesirable, etc.).

Emotions play an important role in signaling the salience of things, directing attention toward what is important and away from distractions, thereby helping to prioritize concerns (Picard, 1997). Alice Isen (2000) has studied the beneficial effects of mild, positive affect on a variety of decision-making processes for medical diagnosis tasks (e.g., facilitating memory retrieval; promoting creativity and flexibility in problem solving; and improving efficiency, organization, and thoroughness in decision making). Negative affect allows us to think in a highly focused way under negative, high-stress situations. Positive affect allows us to think creatively and make broad associations in a relaxed positive state.

Furthermore, whereas too much emotion can hinder intelligent thought and behavior, too little emotion is even more problematic. The importance of emotion in intelligent decision making was demonstrated by Damasio in studies of patients with neurological damage that impaired their emotional systems (Damasio, 1994). Although these patients performed normally on standardized cognitive tasks, their ability to make rational and intelligent decisions in their daily lives was severely limited. For instance, they may have lost a lot of money in an investment, but, instead of becoming more cautious and curtailing investing, these emotionally impaired patients continued to invest. Because they did not seem to link bad feelings and dangerous choices, they continued to make the same choices again and again. The same pattern was repeated in relationships and social interactions, sometimes resulting in the loss of jobs, friends, and so on.

Highly functioning autistics reveal the crucial role of emotion in normal relations. They seem to understand the emotions of others like a computer—they memorize and follow rules to guide their behavior but lack an intuitive understanding of others. In short, they are socially handicapped because they cannot understand or interpret the social cues of others or respond in a socially appropriate way (Baron-Cohen, 1995).

Emotion-inspired mechanisms and capabilities will be essential to the success of autonomous robots. Many more examples could be given to illustrate the importance of social and emotion-inspired mechanisms and abilities to robots that must make decisions in complex and uncertain circumstances, either working alone or with other robots. Our primary interest, however, is how social and emotion-inspired mechanisms can *improve* the way robots function in the human environment and enable them to work effectively in partnership with people.

This does not imply that a robot's emotion-based or cognition-based mechanisms and capabilities must be identical to those in natural systems. The question of whether or not robots can feel human emotions, for example, is irrelevant to our purposes. Furthermore, providing social-based and emotion-based mechanisms should not be glossed over as merely building "happy" or entertaining robots. To do so would be to miss an extremely important point. Just as they do in living creatures, social and emotion-inspired mechanisms can be used to modulate the cognitive systems of the robot to make it function *better* in a complex, unpredictable environment—enabling it to make *better* decisions, to learn more effectively, and to interact more appropriately with others than it could with its cognitive system alone. Therefore, by designing integrated systems for robots with internal mechanisms that complement and modulate their cognitive capabilities with the regulatory, signaling, biasing, and other attention, value assessment, and prioritization mechanisms associated with emotion systems in living creatures, we will effectively be giving robots a system that serves the same useful functions that emotions serve in us—no matter what we call it.

The purpose of this short paper is to put forth an argument for social and emotional intelligence in the design of personal robots that assist and entertain their human users. (For an exploration of these issues from an artistic, scientific, and technological perspective, see Breazeal, 2002). Specific research projects in our laboratory are being conducted on how robots with social-emotive capabilities can assist human astronauts in space, perform opposite human actors in film, and serve as learning companions for children. For more information about this research, see *<http://robotic.media.mit.edu/>*.

REFERENCES

Asimov, I. 1942 (reprinted 1991). I Robot. New York: Bantam Books.

Baron-Cohen, S. 1995. Mindblindness. Cambridge, Mass.: MIT Press.

Bluethmann, W., R. Ambrose, M. Diftler, E. Huber, M. Goza, C. Lovchik, and D. Magruder. 2003. Robonaut: a robot designed to work with humans in space. Autonomous Robots 14(2-3): 179–207.

Braitenberg, V. 1984. Vehicles: Experiments in Synthetic Psychology. Cambridge, Mass.: MIT Press.

Breazeal, C. 2002. Designing Sociable Robots. Cambridge, Mass.: MIT Press.

Damasio, A. 1994. Descartes' Error: Emotion, Reason, and the Human Brain. New York: G.P. Putnam's Sons.

Doyon, A., and L. Liaigre. 1966. Jacques Vaucanson, mécanicien de genie (Jacques Vaucanson, Genius Mechanic). Paris: Presses Universitaires de France.

Fong, T., I. Nourbakhsh, and K. Dautenhahn. 2003. A survey of socially interactive robots. Robotics and Autonomous Systems 42(3-4): 143–166.

Forlizzi, J., C. Di Salvo, and F. Gemperle. 2004. Assistive robotics and an ecology of elders living independently in their homes. Human-Computer Interaction 19: 25–59.

Isen, A. 2000. Positive Affect in Decision Making. Pp. 261-277 in Handbook of Emotions, 2nd ed., edited by M. Lewis and J. Haviland. New York: The Guildford Press.

Kiesler, S., and J. Goetz. 2002. Mental Models of Robotic Assistants. Pp. 576–577 in Proceedings of CHI 2002 Conference on Human Factors in Computing Systems. New York: ACM Press.

Lindsay, D. 1997. Talking head. American Heritage of Invention and Technology 13(1): 57–63.

McCartney, S. 1999. ENIAC: The Triumphs and Tragedies of the World's First Computer. New York: Walker and Company.

Norman, D. 1990. The Design of Everyday Things. New York: Basic Books.

Norman, D. 2004. Emotional Design. New York: Basic Books.

Norman, D., A. Ortony, and D. Russell. 2003. Affect and machine design: lessons from the development of autonomous machines. IBM Systems Journal 41(1): 39–44.

Nourbakhsh, I., J. Bobenage, S. Grange, R. Lutz, R. Meyer, and A. Soto. 1999. An affective mobile robot with a full time job. Artificial Intelligence 114(1-2): 95–124.

Picard, R. 1997. Affective Computation. Cambridge, Mass.: MIT Press.

Reeves, B., and C. Nass. 1996. The Media Equation. Palo Alto, Calif.: CSLI Publications.

Rosheim, M. 1994. Robot Evolution: The Development of Anthrobotics. New York: John Wiley and Sons.

Rosheim, M. 2000. L'automa programmabile di Leonardo. XL Lettura Vinciana, 15 aprile 2000. Citta' di Vinci, Biblioteca Comunale Leonardiana. Florence, Italy: Giunti Gruppo Editoriale.

Sloman, A., and M. Croucher. 1980. Why robots will have emotions. Pp. 197–202 in Proceedings of the 7th International Conference on Artificial Intelligence. Menlo Park, Calif.: International Joint Conferences on Artificial Intelligence.

Thomas, F., and O. Johnson. 1981. The Illusion of Life. New York: Hyperion.
Turing, A.M. 1950. Computing machinery and intelligence. Mind 59(236): 433–460.
UNECE and IFR (United Nations Economic Commission for Europe and the International Federation of Robotics). World Robotics 2002. New York and Geneva: United Nations Publications.
Walter, W.G. 1950. An imitation of life. Scientific American 182: 42–54.

APPENDIXES

Contributors

Adam Paul Arkin is a faculty scientist in the Physical Biosciences Division at E.O. Lawrence Berkeley National Laboratory; an assistant professor of bioengineering at the University of California, Berkeley; and an assistant investigator, Howard Hughes Institute of Medical Research. His research is focused on physical chemistry of the cellular interior, nonlinear and stochastic dynamics, analysis and modeling of cellular processes, analysis of biological data, bioinformatics, biosensors, and genetics and cell biology. Dr. Arkin is the recipient of the *Technology Review* TR100 Award. He received a B.A. from Carleton College and a Ph.D. from the Massachusetts Institute of Technology.

Jon Berkoe is senior principal engineer and manager of the Advanced Simulation and Analysis Group at Bechtel National, Inc., in San Francisco, California. His primary responsibilities include coordinating the deployment of R&D capabilities to Bechtel projects around the world, managing a staff of experienced specialists, and budgeting resources for software and hardware expenditures. With 18 years of mechanical engineering experience as a specialist in fluid dynamics and heat transfer, Mr. Berkoe has applied his expertise in computational fluid dynamics modeling to large engineering projects involving ventilation systems, process equipment, pipelines, metallurgical operations, and environmental flows. Prior to joining Bechtel, he worked for the Spacecraft Division of Lockheed Corporation and the Nuclear Division of General Electric Company.

Mr. Berkoe received B.S. and M.S. degrees in mechanical engineering from the Massachusetts Institute of Technology.

Cynthia Breazeal is an assistant professor of media arts and sciences at the Massachusetts Institute of Technology (MIT), where she directs the MIT Media Laboratory Robotic Life Group and holds the LG Career Development chair. Previously, she was a postdoctoral associate in the MIT Artificial Intelligence Laboratory. Dr. Breazeal's pioneering research is concentrated on the art and science of human-robot interaction and cooperation and the development of robots that can be partners to humans and play a valuable, rewarding, and unprecedented role in the everyday lives of ordinary people. Kismet, her anthropomorphic robotic head, has been featured in international media and is the subject of her book, *Designing Sociable Robots,* published by MIT Press. She continues to develop anthropomorphic robots as part of her ongoing work of building artificial systems that can learn from and interact with people in an intelligent, lifelike, sociable manner. Dr. Breazeal received Sc.D. and M.S. degrees in electrical engineering and computer science from MIT and a B.S. in electrical and computer engineering from the University of California, Santa Barbara.

Greg P. Carman is head of the Active Materials Laboratory in the Mechanical and Aerospace Engineering Department at the University of California, Los Angeles (UCLA). He was chairman of the Adaptive Structures and Material Systems Section of the American Society of Mechanical Engineers (ASME) in 2000-2002, is an associate editor for the *Journal of Intelligent Material Systems Structures,* and is on the editorial advisory board of *Journal of Composite Materials.* He was awarded the Northrop Grumman Young Faculty Award in 1995 for his research at UCLA on active materials and received two Best Paper Awards from the ASME Adaptive Structures and Material Systems Committee in 1996 and 2001. In 2002, he was made honorary professor of the University of Baoutou China, and in 2003, he was elected a Fellow of ASME. In 2004, Dr. Carman was awarded the ASME Adaptive Structures and Material Systems Prize for his contributions to smart materials and structures and his life-long commitment to this field. His main interest is in the basic mechanics and materials issues related to coupled electro-magneto-thermo-mechanical materials. He received a B.S. in engineering science and mechanics from Virginia Polytechnic Institute and State University (VPI), an M.S. in metallurgical and materials engineering from the University of Alabama, Tuscaloosa, and a Ph.D. in engineering mechanics from VPI.

Paul Debevec is a research assistant professor in the Computer Science Department of the University of Southern California (USC) and executive producer of graphics research at the USC Institute for Creative Technologies, where he directs research on the creation of realistic virtual actors and environments. Dr.

Debevec's Ph.D. thesis (UC-Berkeley, 1996) presented an image-based modeling and rendering system for creating photoreal architectural models from photographs. Based on this system, he led a team that created a photorealistic model of the Berkeley campus for his 1997 film, *The Campanile Movie*; the techniques were later used to create the Academy Award-winning virtual backgrounds in the 1999 film, *The Matrix*. Dr. Debevec has developed techniques for capturing real-world illumination and for illuminating synthetic objects with real light, facilitating the integration of real and computer-generated imagery; these techniques were demonstrated in his 1999 film, *Fiat Lux*. He has also led the development of the "Light Stage," a device that allows objects and actors to be synthetically illuminated with any form of lighting, recently used to create photoreal digital actors for the film *Spider-Man 2*. In 2001, he received the first Significant New Researcher Award from the Association for Computing Machinery (ACM) Special Interest Group on Computer Graphics and Interactive Techniques (SIGGRAPH), and in 2002, he received the TR 100 Award from *Technology Review*.

William G. Gardner is the founder and president of Wave Arts, Inc., a company that develops audio signal-processing software, located in Arlington, Massachusetts. Wave Arts sells software to audio-production professionals and licenses algorithms to consumer electronics manufacturers of home audio-visual equipment, mobile devices, and other products. During his years as a student at Massachusetts Institute of Technology (MIT), his research was focused on spatial audio, reverberation, sound synthesis, real-time signal processing, and psychoacoustics. He was awarded a Motorola Fellowship at the Media Laboratory and was the recipient of the 1997 Audio Engineering Society Publications Award for a paper on low-latency convolution. From 1984 to 1990, he worked at Kurzweil Music Systems developing software and signal-processing algorithms for electronic musical instruments. Dr. Gardner received a B.S. in computer science and engineering (1982), an M.S. in media arts and sciences (1992), and a Ph.D. in media arts and sciences (1997), all from MIT. He is a member of the Audio Engineering Society.

B. Kent Joosten is a member of NASA's Exploration Systems Engineering Office at Johnson Space Center in Houston, Texas. From 1980 through 1990, Mr. Joosten served as a flight designer and mission controller for the Space Shuttle; he worked in the Mission Control Center during more than 30 Space Shuttle missions. After the Challenger accident, he participated in the modification of the Shuttle Orbiter guidance and navigation characteristics and the development of flight crew procedures and computer software to improve the Orbiter's survivability. Beginning in 1995, Mr. Joosten managed projects to demonstrate the extraction of oxygen from lunar soil. Since 1997, he has participated in the development of broad-based strategies for future human exploration of the Moon,

Mars, and beyond, including launch, space transportation, and surface exploration strategies, technology planning, demonstration project planning, and robotic mission payload definition. He is a senior member of the American Institute of Aeronautics and Astronautics and the recipient of numerous Group Achievement Awards. Mr. Joosten received a B.S. and M.S. in aerospace engineering from Iowa State University.

Ioannis G. Kevrekidis is a professor in the Department of Chemical Engineering at Princeton University, where he holds concurrent appointments as a faculty member in the Program in Applied and Computational Mathematics and an associated faculty member in the Department of Mathematics. His research is on scientific computation for complex/multiscale systems modeling with an emphasis on nonlinear dynamics. Dr. Kevrekidis earned his undergraduate diploma from the National Technical University of Athens and his M.A. in mathematics and Ph.D. in chemical engineering from the University of Minnesota. He has been a Packard Fellow and a National Science Foundation Presidential Young Investigator, and his work has been recognized by the American Institute of Chemical Engineers (Allan P. Colburn Award) and Society for Industrial and Applied Mathematics (J.D. Crawford Prize).

Leslie A. Momoda is director of the Sensors and Materials Laboratory at HRL Laboratories LLC in Malibu, California. She has 17 years of experience in the field of materials synthesis, processing, and characterization for electronic, thermal, and structural applications. She currently leads a group of 20 engineers and scientists conducting research on active energy storage, sensing, and thermal materials and is personally in charge of several major projects on smart materials, low-temperature processing of ceramics materials, and materials and techniques for thermal management. She is also the principal investigator for a DARPA program on compact hybrid actuation. During her career, Dr. Momoda has conducted research on ionic conduction, crystal growth, microstructural control, and electrical properties of sol-gel-derived thin films (including PZT and BST), latent heat techniques for enhanced heat transfer, and novel active materials. She is the author or coauthor of 17 published papers and the owner of two patents. Dr. Momoda received a B.S. in chemical engineering and an M.S. and Ph.D. in materials science and engineering, all from the University of California, Los Angeles.

Rob Phillips is a professor of mechanical engineering and applied physics at the California Institute of Technology, where his research is focused on nanoscale mechanics in biological systems. He received a B.S. from the University of Minnesota and a Ph.D. from Washington University.

Laura R. Ray is an associate professor of engineering at Dartmouth College. She received a B.S.E., summa cum laude, and a Ph.D. in mechanical and aerospace engineering from Princeton University and an M.S.E. from Stanford University, where she won first prize in the Lincoln National Design Competition for her master's project. The recipient of numerous fellowships and awards, Dr. Ray was a faculty member at Clemson University and Christian Brothers University before joining the Dartmouth Thayer School of Engineering in 1996. Her research has been sponsored by the National Science Foundation, the National Research Council, the Air Force Office of Scientific Research, and a number of local and international companies. The author or coauthor of more than 40 refereed articles and conference publications on robust control, nonlinear estimation and control, active noise control, and applications of control theory to mechanical systems, ground vehicles, and air transportation systems, she is also a member of the American Society of Mechanical Engineers and the Institute of Electrical and Electronics Engineers. She teaches courses in control theory, dynamics, and computer-aided design and analysis.

Tommaso P. Rivellini, the lead mechanical engineer for the Mars Science Laboratory (MSL) Entry, Descent, and Landing Program at the Jet Propulsion Laboratory (JPL) in Pasadena, California, is responsible for the development of the entry, descent, and landing hardware on the MSL Mars lander scheduled for launch in 2009. He has held numerous other positions at JPL, including deputy mechanical-systems architect for the Mars Exploration Rover, mechanical prototype task leader for the second-generation Mars lander, Mars sample return mechanical-system engineer, Deep Space 2 Mars microprobe project element manager, and member of the Mars Pathfinder entry, descent, and landing team; he subsequently was responsible for the design and development of the air-bag subsystem. He has received numerous awards, including the American Institute of Aeronautics and Astronautics Engineer of the Year Award, the Chrysler Award for Innovation in Design, the NASA Exceptional Achievement Medal, and the *Design News* Magazine Excellence in Design Award, all for his work on the Mars Pathfinder air-bag subsystem. Mr. Rivellini received a B.S. in aerospace/ mechanical engineering from Syracuse University and an M.S. in aerospace engineering from the University of Texas at Austin.

Bjorn B. Stevens is an associate professor in the Department of Atmospheric and Oceanic Sciences at the University of California, Los Angeles (UCLA), where he is responsible for a program of research and teaching with a focus on the role of the planetary boundary layer in large-scale circulations. He is also an affiliate scientist at the National Center for Atmospheric Research (NCAR) in Boulder, Colorado, where he is working with the Climate and Global Dynamics and Mesoscale and Microscale Meteorology Divisions to understand and quantify the role of small-scale processes in large-scale circulations. In 1998-1999, he was a visiting scientist with the Max Planck Institut für Meterologie in Hamburg, Germany. Dr. Stevens was a postdoctoral fellow at NCAR from 1996 to 1998. He is recipient of the NASA New Investigator Award, the Clarence Leroy Meisinger Award of the American Meteorological Society, the Editors' Award from the *Journal of Atmospheric Sciences*, a National Science Foundation CAREER Award, and an Alexander von Humboldt Foundation Fellowship. Among his professional activities are, editor, *Journal of Atmospheric Science;* advisory board member, UCLA Academic and Technology Services; organizing committee member, UCLA Center for Computational Sciences and Engineering; reviewer for numerous journals; and member of the American Meteorological Society and the American Geophysical Union. He received B.S. and M.S. degrees in electrical engineering from Iowa State University and a Ph.D. in atmospheric sciences from Colorado State University.

Jennifer L. West is the Isabel C. Cameron Professor in the Departments of Chemical Engineering and Bioengineering at Rice University. She has received the National Science Foundation (NSF) CAREER Award, the Society for Biomaterials Outstanding Young Investigator Award, the Controlled Release Society Cygnus Award for Outstanding Research in Drug Delivery, the Parke-Davis Atorvastatin Award for Research in Applied Vascular Biology, and the *Technology Review* TR100 Award. Her research program is funded by the National Institutes of Health, NSF, U.S. Department of Defense, the state of Texas, and several private industries. Dr. West received a Ph.D. from the University of Texas at Austin.

Program

NATIONAL ACADEMY OF ENGINEERING

Tenth Annual Symposium on
Frontiers of Engineering
September 9-11, 2004

ENGINEERING FOR EXTREME ENVIRONMENTS
Organizers: Mary Kae Lockwood and John W. Weatherly

Cool Robots: Scalable Mobile Robots for Instrument Network Deployment in Polar Climates
Laura R. Ray, Dartmouth College

The Role of Modeling and Simulation in Extreme Engineering Projects
Jon Berkoe, Bechtel National, Inc.

The Challenges of Landing on Mars
Tommaso P. Rivellini, Jet Propulsion Lab

Accessing the Lunar Poles for Human Exploration Missions
B. Kent Joosten, NASA Johnson Space Center

* * *

DESIGNER MATERIALS
Organizers: Kristi S. Anseth and Diann E. Brei

Thin-Film Active Materials
Greg P. Carman, University of California, Los Angeles

139

The Future of Engineering Materials:
Mutlifunction for Performance-Tailored Structures
Leslie A. Momoda, HRL Laboratories, LLC

Biomimetic Strategies in Vascular Tissue Engineering
Jennifer L. West, Rice University

* * *

DINNER SPEAKER

Unlikely Partners: DARPA and Me
Alex Singer, film director

* * *

MULTISCALE MODELING

Organizers: Grant S. Heffelfinger and Dimitrios Maroudas

Equation-Free Modeling For Complex Systems
Ioannis G. Kevrekidis, Princeton University

Modeling the Stuff of the Material World: Do We Need All of the Atoms?
Rob Phillips, California Institute of Technology

Balancing Scales in Biological Models
Adam Paul Arkin, Lawrence Berkeley National Laboratory

Small-Scale Processes and Large-Scale Simulations of the Climate System
Bjorn B. Stevens, University of California, Los Angeles, and
National Center for Atmospheric Research

* * *

ENGINEERING AND ENTERTAINMENT
Organizers: David Baraff and Chris Kyriakakis

*Capturing and Simulating Physically Accurate Illumination
in Computer Graphics*
Paul Debevec, University of Southern California

Spatial Audio Reproduction: Toward Individualized Binaural Sound
William G. Gardner, Wave Arts, Inc.

Designing Socially Intelligent Robots
Cynthia Breazeal, Massachusetts Institute of Technology

Participants

NATIONAL ACADEMY OF ENGINEERING

Tenth Annual Symposium on
Frontiers of Engineering
September 9-11, 2004

Cameron F. Abrams
Assistant Professor
Department of Chemical Engineering
Drexel University

Andrew Alleyne
Ralph and Catherine Fisher Professor
  of Engineering
Department of Mechanical and
  Industrial Engineering
University of Illinois, Urbana-
  Champaign

Luis A. Nunes Amaral
Associate Professor
Department of Chemical and
  Biological Engineering
Northwestern University

Guillermo Ameer
Assistant Professor
Biomedical Engineering Department
Northwestern University

Arnon Amir
Research Staff Member
IBM Almaden Research Center

Kristi S. Anseth
Tisone Professor of Chemical and
  Biological Engineering,
  Associate Professor of Surgery,
  and HHMI Assistant Investigator
Department of Chemical and
  Biological Engineering
University of Colorado, Boulder

Adam P. Arkin
Faculty Scientist, Physical Sciences
  Division,
E.O. Lawrence Berkeley National
  Laboratory
Assistant Professor, Departments of
  Bioengineering and Chemistry
University of California, Berkeley

David Baraff *(unable to attend)*
Senior Animation Scientist
Pixar Animation Studios

Hilary Bart-Smith
Assistant Professor
Department of Mechanical and
    Aerospace Engineering
University of Virginia

Amy E. Bell
Assistant Professor
Electrical and Computer Engineering
Virginia Polytechnic Institute and
    State University

Jon Berkoe
Senior Principal Engineer and
    Manager
Advanced Simulation and Analysis
Bechtel National, Inc.

M. Brian Blake
Assistant Professor
Department of Computer Science
Georgetown University

Lawrence E. Bool
Senior Development Associate
Industrial Applications and
    Healthcare R&D
Praxair, Inc.

Cynthia Breazeal
Assistant Professor of Media Arts and
    Sciences, LG Career
    Development Professor of Media
    Arts and Sciences, and Director
    of Robotic Life Group
Massachusetts Institute of Technology

Diann Brei
Associate Professor
Department of Mechanical
    Engineering
University of Michigan

Mark D. Breyen
Principal Scientist
Medtronic, Inc.

Greg P. Carman
Professor
Department of Mechanical and
    Aerospace Engineering
University of California, Los Angeles

John E. Carsley
Research Engineer
General Motors Research and
    Development Center

Danielle Chamberlin
Member of Technical Staff
Micro and Nanoscale Technologies
    Department
Agilent Laboratories

Christina Chan
Associate Professor
Chemical Engineering and Materials
    Science Department
Michigan State University

Eric F. Charlton
Engineering Staff
Lockheed Martin Aeronautics Co.

Elaine Chew
Assistant Professor
Daniel J. Epstein Department of
    Industrial and Systems
    Engineering
University of Southern California

Paul S. Chinnock
Principle Systems Engineer
Systems Engineering
Raytheon Missile Systems

Robert L. Clark
Thomas Lord Professor
Department of Mechanical
    Engineering and Materials
    Science
Duke University

Tracy Clarke-Pringle
Senior Consulting Engineer
DuPont Engineering Technology
DuPont Company

Pablo G. Debenedetti
Class of 1950 Professor
Department of Chemical Engineering
Princeton University

Paul Debevec
Executive Producer, Graphics
    Research, and Research Assistant
    Professor
Institute for Creative Technologies
University of Southern California

Bryan G. Dods
Manager
Assembly and Quality Technology
Boeing Company

Robert J. Drost
Principal Research Scientist
Sun Microsystems Research
    Laboratories

John Dunagan
Researcher
Microsoft Research

Ziyad H. Duron
Jude and Eileen Laspa Professor of
    Engineering
Harvey Mudd College

David B. Fogel
Chief Executive Officer
Natural Selection, Inc.

David M. Ford
Associate Professor
Department of Chemical Engineering
Texas A&M University

Daniel R. Gamota
Senior Manager and Distinguished
    Member of the Technical Staff
Motorola Corporation

William G. Gardner
Founder and President
Wave Arts, Inc.

James L. Garrison
Assistant Professor
School of Aeronautics and
    Astronautics
Purdue University

R. Brent Gillespie
Assistant Professor
Department of Mechanical
    Engineering
University of Michigan

Ronald Grundbacher
Section Head
HEMT Technology
Northrop Grumman Space
    Technology

Justin Hanes
Assistant Professor
Department of Chemical and
    Biomolecular Engineering
John Hopkins University

Babak Hassibi
Associate Professor
Department of Electrical Engineering
California Institute of Technology

Grant S. Heffelfinger
Deputy Director
Materials and Process Sciences
    Center
Sandia National Laboratories

Markus Heinimann
Technical Specialist - Aerospace
Product Design and Analysis
Alcoa Technical Center

Markus Hofmann
Director
Services Infrastructure Department
Bell Laboratories, Lucent
    Technologies

Ayanna Howard
Senior Robotic Researcher
Jet Propulsion Laboratory

Nancy L. Johnson
Lab Group Manager
General Motors Research and
    Development Center

B. Kent Joosten
Deputy Manager
Exploration Analysis and Integration
    Office
NASA Lyndon B. Johnson Space
    Center

Jeyhan Karaoguz
Senior Principal Scientist
Broadcom Corporation

Ioannis G. Kevrekidis
Professor
Department of Chemical Engineering
Princeton University

Chris Kyriakakis
Associate Professor, Department of
    Electrical Engineering
Research Area Director, Sensory
    Interfaces
University of Southern California

Mihir Lal
Senior Research Scientist
MicroCoating Technologies, Inc.

Kristin Lauter
Researcher
Microsoft Research

Erin Lavik
Assistant Professor
Department of Biomedical
    Engineering
Yale University

Kelvin H. Lee
Associate Professor
School of Chemical and Biomolecular
    Engineering
Cornell University

Sophia Lefantzi
LT Technical Staff
Combustion Research Facility
Sandia National Laboratories

Valerie J. Leppert
Assistant Professor of Materials
    Engineering
Division of Engineering
University of California, Merced

Mary Kae Lockwood
Aerospace Engineer
NASA Langley Research Center

Yueh-Lin Loo
Assistant Professor
Department of Chemical Engineering
University of Texas, Austin

Cristina Videira Lopes
Assistant Professor
School of Information and Computer
    Science
University of California, Irvine

Garrick E. Louis
Associate Professor
Department of Systems and
    Information Engineering
University of Virginia

James Maher
Chief Technical Advisor
Technip Offshore, Inc.

Dimitrios Maroudas
Professor
Department of Chemical Engineering
University of Massachusetts

Paul F. McKenzie
Executive Director
Process R&D and Technology
    Transfer
Bristol Myers Squibb

Leslie A. Momoda
Director
Sensors and Materials Laboratory
HRL Laboratories, LLC

Michael A. Mooney
Associate Professor
Division of Engineering
Colorado School of Mines

John F. Muth
Assistant Professor
Department of Electrical and
    Computer Engineering
North Carolina State University

Hock M. Ng
Member of Technical Staff
Bell Laboratories, Lucent
    Technologies

Jason Norman
Section Manager
PGS Technology
BOC Group

Laura O'Donnell
Owner
Synthese, LLC

Avinash S. Patwardhan
Principal Water Resources Engineer
CH2M Hill

Rob Phillips
Professor of Mechanical Engineering
    and Applied Physics
Department of Mechanical
    Engineering
California Institute of Technology

Shawn H. Phillips
Chief
Material Applications Branch,
    Propulsion Directorate
Air Force Research Laboratory

Patrick Piccione
Research Scientist
Strategic Research
ATOFINA Chemicals, Inc.

David Polidori
Director
Research and Development
Entelos, Inc.

Vikas Prakash
Professor
Department of Mechanical and
    Aerospace Engineering
Case Western Reserve University

David Putnam
Assistant Professor
School of Chemical and Biomolecular
    Engineering
Cornell University

Raj Rajiyah
Executive Director
Applied Mechanics and Materials
    Engineering
Cummins, Inc.

Sujatha Ramanujan
Research Scientist
Eastman Kodak Company

Ainissa G. Ramirez
Assistant Professor
Department of Mechanical
    Engineering
Yale University

Laura R. Ray
Associate Professor
Thayer School of Engineering
Dartmouth College

Tommaso P. Rivellini
Principal Mechanical Engineer
Jet Propulsion Laboratory

Virginia Robbins
Principal Project Scientist
Agilent Technologies

Noble L. Rye
Engineer
Modeling Simulation and Analysis
Procter & Gamble Corporation

Gerwin Schalk
Research Scientist III/Chief Software
    Engineer
New York State Department of
    Health

Scott E. Schoenfeld
Chief Scientist
Impact Physics
Army Research Laboratory

Sharon Settnek
Program Manager
Science Applications International
    Corporation

Yang Shao-Horn
Assistant Professor
Department of Mechanical
    Engineering
Massachusetts Institute of Technology

Kendra V. Sharp
Assistant Professor
Department of Mechanical and
    Nuclear Engineering
Pennsylvania State University

Eric V. Shusta
Assistant Professor
Department of Chemical and
    Biological Engineering
University of Wisconsin-Madison

Karen R. Smilowitz
Associate Professor
Department of Industrial Engineering
    and Management Sciences
Northwestern University

Jonathan M. Smith
Program Manager
Defense Advanced Research Projects
    Agency

Christopher L. Soles
Materials Research Engineer
Polymers Division, Electronics
    Applications Group
National Institute of Standards and
    Technology

Donald W. Stanton
Technical Advisor
Combustion Research
Cummins, Inc.

Bjorn B. Stevens
Associate Professor
Department of Atmospheric and
    Oceanic Sciences
Universiy of California, Los Angeles

Natacha Supper
Engineer/Scientist
Hitachi Global Storage Technology
    Research Center

Julie Swann
Assistant Professor
School of Industrial and Systems
    Engineering
Georgia Institute of Technology

Albert S. Tam
Research Manager
DuPont Engineering

Eric G. Thaler
Technology Manager
GE Advanced Materials

John L. Uhrie
Superintendente de Proyectos
Phelps Dodge Mining Corporation

Srinivas Veeramasuneni
Senior Member Technical Staff
USG Research and Technology
    Center
USG Corporation

Ramesh Visvanathan
Department Head
Real-time Vision and Modeling
Siemens Corporate Research, Inc.

John M. Vohs
Carl V. Patterson Professor and Chair,
    Department of Chemical and
    Biomolecular Engineering
University of Pennsylvania

Joseph J. Wang
Associate Professor
Department of Aerospace and Ocean
    Engineering
Virginia Polytechnic Institute and
    State University

Yong Wang
Chief Scientist and Team Leader
Catalysis and Reaction Engineering
Pacific Northwest National
     Laboratory

Clement Wann
Senior Manager
IBM Semiconductor Research and
     Development Center

Sunil G. Warrier
Project Leader
Solid Oxide Fuel Cell Program
United Technologies Research Center

Gregory Washington
Associate Professor
Department of Mechanical
     Engineering
Ohio State University

John W. Weatherly
Ice Geophysicist
U.S. Army Cold Regions Research
     and Engineering Laboratory

Jennifer L. West
Isabel C. Cameron Professor of
     Bioengineering and Associate
     Professor of Chemical
     Engineering
Departments of Chemical and
     Bioengineering
Rice University

Ming-Cheng Wu
Staff Research Scientist
Delphi Corporation

Xiang Zhang
Professor and Director
NSF Nanoscale Science and
     Engineering Center
University of California, Los Angeles

Dinner Speaker

Alex Singer
Film Director

Guests

Jennifer Cho
Senior Product Development Manager
Publications Division
American Chemical Society

Kenneth E. Harwell
Senior Science and Technology
     Advisor to the Deputy
     Undersecretary of Defense for
     Laboratories and Basic Sciences
Department of Defense

Anne Matsuura
Program Manager
Electronics and Physics Directorate
Air Force Office of Scientific
     Research

Carey Schwartz
Program Manager
Defense Sciences Office
Defense Advanced Research Projects
     Agency

Judith Singer
Guest of Alex Singer

Doug Stetson
Deputy Director
Advanced Planning and Integration
    Office
National Aeronautics and Space
    Administration

Sarah Tegen
Editorial Associate
The National Academies

National Academy of Engineering

W. Dale Compton
Home Secretary

Lance A. Davis
Executive Officer

Donna Dean
Senior Scholar-in-Residence

Janet Hunziker
Program Officer

Jenny Hardesty
Senior Project Assistant